T0136120

AN ERRANT EYE

AN

Errant

EYE

Poetry and Topography
in Early Modern France

Tom Conley

University of Minnesota Press
Minneapolis
London

Published by the University of Minnesota Press
111 Third Avenue South, Suite 290
Minneapolis, MN 55401-2520
http://www.upress.umn.edu

Library of Congress Cataloging-in-Publication Data

Conley, Tom.
    An errant eye : poetry and topography in early modern France / Tom Conley.
        p.   cm.
    Includes bibliographical references and index.
    ISBN 978-0-8166-6964-6 (hc : alk. paper) — ISBN 978-0-8166-6965-3
(pbk. : alk. paper)
    1. French poetry—16th century—History and criticism. 2. Geography
in literature. 3. Cartography in literature. 4. Space in literature. 5. Place
(philosophy) in literature. 6. Cartography—France—History—16th century.
I. Title.
    PQ418.C66 2010
    841'.30932—dc22

                                                            2010041232

Printed in the United States of America on acid-free paper

The University of Minnesota is an equal-opportunity educator and employer.

17 16 15 14 13 12 11   10 9 8 7 6 5 4 3 2 1

Contents

Preface and Acknowledgments

In a recent and wondrous study of maps of France and its regions from the incunabular age of Ptolemy to the end of the sixteenth century, Monique Pelletier (2009) notes that cartographers were as much artists as they were surveyors, strategists, and scientists. No less, if they were not heralded among printers, poets, and writers, many of their maps found strong affiliation with a growing and diverse body of literature of protean shape and form. Maps, paintings, engravings, and poems often found themselves in such creative confluence that a manual of cosmography or a topographical projection could be read with the creative urgency required of a book of poetry, a piece of satirical narrative, or a personal essay. The chapters that follow, in which poetry is broadly defined to include prose and verse alike, are built on this observation. It presupposes common ties among fields that seem distinct from one another in our time but that were not viewed as separate in theirs. Or, as Pelletier, summing up her review of topographical images from the middle years of the sixteenth century, states when casting her gaze forward from then up to now, "A choice must be made if one is to be a painter and/or a cartographer, but in the sixteenth century it is not yet a necessity" (62).

For this and other reasons this study owes much to a variety of people and institutions of different style and substance. The project was inspired, first and foremost, by the encouragement of the late David

Woodward, who had asked me to develop a section of *The History of Cartography: The European Renaissance* (2007) on literature and cartography, a field of study that had been gaining momentum over the decades including and following the Columbian quincentenary. The confidence that our deeply missed friend had invested in the project has driven it forward. And many who collaborated with him count among those to whom this work is indebted: in France, Jean-Marc Besse, Frank Lestringant, and Monique Pelletier, whose impeccable scholarship on the history of cartography has been a model and a mold; in North America, James Akerman, David Buisseret, and Robert Karrow, friends of the Newberry Library, have been constant guides. J. Theodore Cachey Jr., Matthew Edney, Franco Farinelli, Elisabeth Hodges, Giorgio Mangani, Louisa Mackenzie, Ricardo Padrón, Simone Pinet, Neil Safier, Henry Turner, and Philip Usher brought critical exchange in areas where cartography and literature are of the same measure. I would like, no less, to thank a group of outstanding art historians who offered equal inspiration. I owe very much to Henri Zerner, friend and colleague, with whom I have had the privilege and pleasure of coteaching an ongoing seminar on text and image in sixteenth-century France. Michael Gaudio, Larry Silver, and Christopher Wood have been points of reference for their telling observations on the printed image and the fortunes of its replication.

Colleagues in early modern literature have been mentors. The work of François Rigolot on poetry and motion stands high over this project. His friendship and example are outstanding. I very much thank Jean-Claude Carron, Juliette Cherbuliez, and Timothy Murray, who carefully read and suggested alterations of the manuscript. Cécile Alduy, Deborah Lesko Baker, Terence Cave, Richard Cooper, Philippe Desan, Michael Giordano, Hope Glidden, Michel Jeanneret, Lawrence Kritzman, Ullrich Lauger, Mary McKinley, Malcolm Quainton, Michael Randall, Richard Regosin, Antónia Szabari, Richard Tinguely, Colette Winn, and others of the endangered species of *seiziémiste* have been in my thoughts in the course of the writing. As the book was moving ahead, Gérard Defaux brought untold inspiration through his friendship, his passion, and, above all, his abiding commitment to early modern studies. We grieve the loss of his person and his presence.

Research was begun under the generous auspices of a Guggenheim Fellowship that allowed protracted study in the Pusey and Houghton libraries of Harvard University. There David Cobb and Joseph Garver guided research on cartography and Susan Halpert offered extraordinary assistance in research on the illustrated book. Mary Haegert attended to illustrations, and along the way the deeply regretted Thomas Ford brought expertise and real encouragement when the work was in its early stages.

Various chapters were essayed at institutions that generously offered counsel and critical dialogue. I thank students and colleagues at the universities of Oregon, Washington, and Kansas for their patience and interest; Jeremie Korta, who read a first draft; Cristina Capineri, of the University of Florence, and Clara Copeta, of the University of Bari, who offered productive stages for exchange; Dominique de Courcelles, of the École Nationale des Chartes, where an early version of the introduction was put forward.

Colleagues at the University of Minnesota Press generously sustained the research and writing on literature and cartography from its beginnings, with the publication in 1996 of *The Self-Made Map: Cartographic Writing in Early Modern France,* which might now be thought of as one of the panels of a diptych, on which *An Errant Eye* is hinged. I thank Douglas Armato and Richard Morrison, who have had faith in the work throughout its development, and Kristian Tvedten, who brought exacting attention to the manuscript. Mary Byers contributed a meticulous hand to the copyediting, for which, as with other projects at the University of Minnesota Press, I am deeply grateful.

It is to Verena Conley, who patiently witnessed this work in its slow growth over the course of too many years, that I dedicate this book.

Introduction: A Snail's Eye

Lost in the labyrinth, adrift in the ruins of a palace of memory, the hero of Francesco Colonna's *Songe de Poliphile* (the French translation, dated 1546, of the *Hypnerotomachia poliphilii* of 1499), a lavishly illustrated novel of quest, erotic venture, and pagan initiation, wanders about without resolve. Groping in darkness, he does not know how to proceed or where to go until it dawns on him that he ought to become a snail to better touch the things in his midst:

> Et combien que mes yeux se trouuassent aucunement accoustumez a ces tene-bres, toutesfois ie ne pouoie rien veoir: parquoy falloit que mes bras feissent l'office de mes yeux, aussi bien que le Lymacon, qui va tastant le chemin avec ses cornes, & s'il treuue empeschement, les retire soudain a soy. En tele maniere i'alloie tastonnant atrauers ces destours aueuglez, & par ces sentes desuoiees, en plus grand travail & perplexité, que Mercure quand il se feit Cigogne.

> [And no matter how much my eyes seemed accustomed to these shadows, I nonetheless couldn't see anything. That is why my arms had to perform my eyes' duties, much like the snail that goes about feeling *[tastant]* its way with its horns, and when he meets an obstacle, he draws them back into himself. It was in that manner I went groping, by trial and error, along those blind detours and paths leading off the beaten track, with greater labor and perplexity than Mercury when he became a stork.][1]

The narrator's words about his ocular touch are set here to serve as an epigraph to this study. The hero finds himself in a nightmare that is rich

in special effects in both its moment and ours. Lost in a cave that re-
minded him of the dark dwelling of Cyclops, the one-eyed monster or
the cavern concealing the evil Cacus, he recalls how bats buzzed about
his ears and how he was so consumed by fright that he thought he was
hearing, feeling, and touching a "cruel dragon"—the very beast seen in
a woodcut on the opposite folio, arched on a rock, spewing and spitting
in front of the helpless wanderer who flees into a dark hallway in the
background. When he thinks of himself as a snail, Poliphile regains his
wits and finds his way through the strange world he has entered. And
when we see him portrayed in the decor shown in the image opposite
his words (fol. 19v), we also find our bearings in the welter of images
and printed forms of the copiously illustrated novel.

The aim of this work is one of retrieving the snail's sense, much like
that which allows Colonna's narrator to gain a compass on the locale
in which he moves. The touch and tact that he acquires when he "becomes snail" are much like the traits of the poet who attends to detail
or the surveyor who gets the lay of the land for which he will draw a
topographical image. His errant eye that "goes about feeling its way" is
an emblem of the ocular character of the reflections that follow. A first
guiding line is that in the age of discovery in Renaissance France—
dating roughly from the spread of the news of the Columbian discoveries and the circulation of Amerigo Vespucci's letters describing what
the Florentine traveler had seen of the New World in the early 1500s up
to the moment, in 1580, when Montaigne publishes his celebrated essay
"Des cannibales"—the eye wanders about and literally *touches* a world
of unforeseen expanse. Unlike the binocular vision of the modern age,
which is said to seek and control the world by virtue of the distance it
measures when looking down and upon it, the errant eye of the sixteenth century is enmeshed and, much like Poliphile, often lost in its
milieu. Endowed with a sense of tactility and, now and again, another
of tact, it moves forward and backward, now alert and then withdrawn,
about and around its ambient world.

The haptic eye becomes a point of reference for poets and artists who
look closely at their territory, who discern its size and scale; who carefully draw the lines defining the nooks, edges, and crannies and relief
of its surface; who follow the roads and rivers cutting and winding
through it; and no less, like the hero of the *Songe*, who touch the lairs
and recesses inspiring the fantasies that mark what they see and what,
in turn, we observe before our eyes. The snail becomes the emblem, too,

of the topographer who records images of regions that seem somehow to be part of a greater whole, a *mundus,* whose limits and even whose character are cast in doubt. At a moment, perhaps inverse to our own, when increased proximity of formerly distant places causes us to wonder less and less about the difference between things here and there or local and global, we can ask ourselves if still today, as it was defined in early manuals of cosmography, the relation of topography to geography retains a shard of validity. The response in the pages to follow will be in the affirmative. When we pause to wonder both where we think we are and how and why, we begin to look at the world from the topographer's point of view: from a position that is both of the sentient body and detached from it, that needs to displace itself to obtain its bearings. In other words, it must wander sometimes with purpose, sometimes aimlessly, but for the most part like the snail, whose eyes touch the world in which it moves.

A second line of inquiry is that the poet's vision is much like that of the topographer who sees, discerns, and orders the world in consort with the art of illustration. The poet is like a cartographer insofar as it is his or her task to describe the world by mixing images, visual designs, and both aural and optical traits of language. Both the mapmaker and the poet often collaborate with artists and printers who attend to the choice of character, to the disposition of printed forms, and to the aspect of images that accompany verse and prose. They are guided to some degree, at once consciously and unconsciously, by a will to reach out to welcome the ambient world and, no sooner perhaps, to admire its splendor prior to using language to gain a contemplative hold upon what they will describe. In this respect historians of the representation of space might be eager to relate the topographer's descriptions to a nascent colonial sensibility while others, adepts of philosophies of vision, readily associate them with a medieval heritage in which sight and touch are taken to be of the same order.[2] Both anthropologists and historians of religion have shown that the act (and art) of *naming* what is felt and discovered in an unknown relation submits the object to control. It is a truism that to name is to colonize, but it is less so to recall that the act follows and often represses a moment of raptness and of marvel, because, unknown, what is seen and touched remains vital to life.[3]

The itinerary that follows leans toward the precolonial moment of the gaze, that which comes before the name, for the purpose of displacing into our own world a sense of wonder and of contemplation that we feel belong to the poet and the topographer: to slow it down, to let its past be touched, but above all to rediscover the transport of the unknown in our milieu through the authors and artists witnessed in the years following the Columbian encounters. It will be imagined that for each of them a wandering eye revels in touching the world much as now, the reader's finger feels the relief of a printed page when it dares (under electronic televisual eyes on the walls of reading rooms and archives) to extend itself to graze the printed page in order to discern the weight of the paper, to esteem the force of an impression left by a woodblock or a line of movable type, or even to follow an inked annotation, alteration, or smudge of a previous hand.

An Event

It often appears that early modern writers and draftsmen use their media to fashion, in the philosophical sense of the term, a continuum of *events*. They create forms in which their own invention and construction, indeed the very perceptions and sensation that give rise to them, are folded into their execution and remain integral to their effects. Philosophers of the baroque age have noted now and again that an event is of a highly tactile nature.[4] It is a moment (if it can be measured in chronological time, which may be indeed unlikely) in which sensation and perception are of the same order, and in which, like the etymology of *invention,* it is something that happens . . . or that the artist causes to happen. An event is what creates a sense of space and place because it calls for a heightened consciousness of *prehension,* that is, of contact with the ambient world in such a way that unforeseen or hitherto unknown relations are made from the experience of things seen and heard in ways that bring forward an uncommon and often passing sense of where and what we are in the world. It is not to say that the event can be associated with the sublime; rather, it is what in the relation of the perceiving organ with the ambient world precipitates an intensity of sensation and perception.

Events can be said to be what yield the perception of space and of how that perception takes "place." If, in the age when the idiolect of semiotics had been integral to critical consciousness, space had been called the "discursive practice of place" (Certeau 1990), the art of reading and seeing could be understood along a similar line, which is that of the third aim of what follows. The design is to let events take place in the areas between verbal and visual forms that belong to a common ground of poetry and cartography. It often seems that the great poems of the French Renaissance "explode" into form and that, likewise, before the viewer's eyes the representation of places in books and atlases extends to include the time and space of their perception. In this way correlated maps and poems continue to be "events" of an order in which their sensate qualities precede the languages that make them communicable.

At the end of "De l'exercitation," the sixth chapter of the second volume of his *Essais*, Montaigne sums up why he has taken the trouble of putting to paper the otherwise harrowing story of a singular event, an unforeseen encounter with death. Every reader of the *Essais* recalls how he tells of going out, not as a knight errant, but merely to go for a ride at a distance of about one league *(une lieuë)* from his chateau. Upon his return a burly friend on a sturdy horse knocks him out of his saddle. Montaigne is thrown to the ground, where, stunned, he lies inert, unconscious, dead to the world around him. He sees and perceives his companions who attend to him and carry him home but is unable to convey any sign of life to them. His body, spattered with clotted blood, is lifted off the ground and carried to safety. Much of the essay that follows depicts the slow and painful return to life that resembles birth and a shattering entry into the world as much as an exit from it.[5] The departure from the chateau is followed by an excruciating return to life. The moment in which the essayist takes an inventory of what may be the net worth of the tale is rich in topographical inflection:

> Ce conte d'un évenement si legier est assez vain, n'estoit l'instruction que j'en ay tirée pour moy; car, à la verité, pour s'aprivoiser à la mort, je trouve qu'il n'y a que de s'en avoisiner. Or, comme dict Pline, chacun est à soy-mesmes une très-bonne discipline, pourveu qu'il ait la suffisance de s'espier de près. Ce n'est pas ci ma doctrine, c'est mon estude; et n'est pas la leçon d'autruy, c'est la mienne.[6]

> [This tale of such a slight event would be quite vain, were it not for the instruction that I have drawn from it for myself; for, in truth, if we are to train ourselves for death, I find that we need only put ourselves in its proximity. Thus, as Pliny says, each is to oneself a very keen discipline, provided that one has vanity enough to spy upon oneself closely. This is not my doctrine, it's my study; and it's not the lesson of others, it's my own.]

Can the gravity of Montaigne's fall from his horse and his brush with death be called *un évenement legier,* a "slight event" of such little consequence that it merits a mere passing mention? In the poetic and spatial arena of the *Essais* the encounter surely can because an event is at once total and local, consuming and contingent, and at once familiar and strange. The classical topic or site of deliberative meditation, "that to philosophize is to learn to die," which had been rehearsed earlier in the *Essais* under a chapter title of the same wording, is here told as an anecdote or an "account" (a *conte* and a *compte*) set in the locale of Gascony. And in the poetic gist of the essay the way to sense or to prehend the end of one's life is at once touched and seen in the surface tension of the verbs of domestication and approximation: *s'apprivoiser . . . s'avoisiner.* To "learn how" to die requires training through the art of writing so that the words themselves can resemble toponyms or even sighting points in the greater landscape of the printed page. When Montaigne notes that everyone ought to cultivate the very good *discipline* of miming death, he refers to Pliny to sustain his argument: "Or, comme dict Pline, chacun est à soy-mesmes une très-bonne discipline." The compression of the name of the classical historian of nature and the nascent art or discipline of self-study makes clear that the *event* of the essay takes place where words rhyme by the way they are seen and heard in the same breath.

In the paronomasia of "what Pliny says here" and *discipline,* Montaigne implies that to spy on oneself can be likened to an autopsy, if not even of a project to map the self in words and in their surrounding space. The event becomes topographical insofar as it opens onto the intimate immensity of his home (which the end of the essay soon confirms) in which he becomes the topic of his own anatomy lesson. By all means he needs vanity enough to spy upon himself from close proximity. In the arcane turns of the *Essais,* in their own idiolect that the author calls his "dictionnaire tout à part moy" (1091/1111), the field that is near *(près)* and around the chateau carries a spatial charge. The ground of the printed page becomes a topographic field on which plot points are marked in graphemes. Among them imaginary lines can be drawn to survey and hence to read the essay visually, at once in its flatness and in its mountainous relief so that it can be treated, much as a work of military cartography, as an orography or a map of various intensities. Montaigne burrows into the folds of the inner self in the extension or *alongeail* he immediately appends to this passage that was scribbled in the lower margin of the Bordeaux copy of 1588.[7] In the formula "suffisance de s'espier de près," a beguiling sibilance draws attention to the visual form of the printed characters. In his memory of the event Montaigne recalls that his horse had been felled and knocked unconscious, "moy dix ou douze pas au delà, mort, estendu à la renverse, le visage tout meurtry et

tout escorché, mon espée que j'avoy à la main, à plus de dix pas au-delà"
(353/373); [myself, ten or twelve steps beyond, dead, thrown, my face all
bruised and skinned, my sword that I had held in my hand, more than
ten steps beyond]. The uncanny echo of Maurice Blanchot's compelling
reflection on death, in *Le pas au-delà* (1973; published in English as *The
Step Not Beyond*), is heard across four centuries of French literature: a
relation with death in the reiterated measure of *à plus de dix pas au-
delà*, in twice "going beyond," localizes the reflection to the degree
that the substantive of the essayist's protective arm, his sword, is shot
through the later description of self-espionage. The *espée* becomes a sort
of pointer or surveyor's theodolite that allows the eye of the essayist to
turn inward and follow a road—*ce chemin* (358/378)—that leads into
the recesses of the self. One of the most famous sentences in all of the
Essais makes clear that the discovery of self-inspection is a particularly
topographical event: "C'est une espineuse entreprinse, et plus qu'il ne
semble, de suyvre une alleure si vagabonde que celle de nostre esprit;
de penetrer les profondeurs opaques de ses replis internes; de choisir
et arrester tant de menus airs de ses agitations" (358/378). [It's a thorny
enterprise, and more than it seems, to follow an allure so wandering as
that of our mind; to penetrate to the opaque depths of its inner folds; to
choose and to arrest so many of the slightest vibrations of its agitation.]
The sword transmutes into the thorny, spiny, or spiked quality of the
espineuse entreprinse in which topography, anatomy, and autopsy are of
the same order.

The sentences bear the sign of an event inasmuch as events happen
and are not prescriptively fashioned. What vibrates or indeed what exceeds the description is the event that opens and abolishes and, in the
same measure, *exhausts*—if the words of a contemporary philosopher are
appropriate—the very space it creates.[8] In *Le pli: Leibniz et le baroque*,
in a chapter whose title poses the question "What is an event?" Gilles
Deleuze proposes an answer by noting that it is a *nexus of prehensions*
by which a prehending agent—that can be Polia of the *Songe* noted earlier or, as we shall see, the snail of the *Hécatomgraphie*—discovers that
it is simultaneously prehended, and vice versa, and with such shimmer
that a continual process of simultaneously inner and outer subjectivation
and objectivation takes place.[9] The self is "othered" in grasping what and
where it is, just as it realizes that the place it apprehends is also exactly
what is apprehending it. The process is one of continuous folding and
refolding, where, in the case—indeed, the fall—of Montaigne the

penetration of the "opaque depths" of the "inner pleats" of the soul involves prehending and discerning surfaces that in turn locate, like an unfolded map, he or she who prehends. Such is an event that bears philosophical and topographical latency, especially where the essayist notes that when he describes himself—and the art of description is at the core of the topographer's métier—he must *sortir en place* (358/378). The formula is multilocal, referring both to a site *chez soi* and to any place wherever. It is evocative of *two sites in one* that displace the very position whence he writes. He is displaced in the very space of his writing. "Il n'est description pareille en difficulté à la description de soy-mesmes, ny certes en utilité. Encore se fault-il testoner, encore se faut-il ordonner et renger *pour sortir en place*" (358/378, emphasis added). [No description is of a difficulty equal to the description of oneself, nor indeed in usefulness. Still one must adorn oneself, be well disposed and neatly kempt to go out in place.]

If Montaigne goes out from his home, his *moyeu,* as he had from his chateau in Gascony, now, in the château (or, in the book itself, in its very place), he goes out with his proper effects and in suitable attire, in the sense of being both *in* and *out,* in a locution that turns the reference to the Gascon landscape into an almost boundless space. The writing crafts a locale and demarcates implied perimeters that in the printed form the words exceed. At the very locus where legions of critics have astutely shown that the project of self-portraiture is born, so also is that of a poetic topography. To describe and figure—*pingere, fingere*—a map or picture of one's being and its place in the world at large suggests that when Montaigne portrays the totality of himself he can only proceed through synecdoche and syllepsis. As in a manual of cosmography, such as that of Pieter Apian, taken up later here, *para pro tota* applies to details that concern the self, much like the eye and the ear, having been detached from the body to illustrate the nature of topography (see Figures 2 and 3), will be put back where they belong, remembered, or indeed situated in a physical and a geographical context.

Here and elsewhere the *Essais* refuse to accede to an Icarian view of a total world while they also entertain its idea or hold to an older—and very modern—concept of a *tout ouvert,* a whole always open, pliable, and of quasi-infinite extension.[10] When elsewhere the author recalls Ptolemy, Apian's model, he remarks that his world vision was erroneously closed, and that it is no more likely that "ce grand corps que nous appellons le monde, est chose bien autre que nous ne jugeons" (555/572) [this great body that we call the world is something quite other than what we judge it to be]; and, further, at the end of the "Apologie de Raimond Sebond," he asserts vehemently that it is impossible to go beyond the severe limits of human measure: "Car de faire la poignée plus grande que le poing, la brassée plus grande que le bras, et d'esperer enjamber plus que de l'estanduë de nos jambes, cela est impossible et monstrueux. Ny que l'homme se monte au dessus de soy et de l'humanité: car il ne peut voir que de ses yeux, ny saisir que de ses prises" (588/604). [For it is impossible and monstrous to make a fistful larger than the fist, an armful larger than the arms, and to hope to stretch our legs wider than they can extend. Nor that man puts himself over and above himself and humanity; for he can see only with his eyes and grasp only what he can hold.] Montaigne's topography reaches far beyond the "place" it takes such pain to describe.

Tact and Sight

The traumatic event that is told in "De l'exercitation" indicates that self-study is engineered through topography and poetry. Its force of expression owes much to its ocular touch and to its acute awareness of space and place. The mental structure on which it is based is clearly drawn in two complementary images that serve as additional epigraphs or, better, as points of reference for the readings that follow. The first (Figure 1) is found in the twentieth emblem of Gilles Corrozet's *Hécatomgraphie* (1541), a vernacular variant on Andreas Alciati's immensely successful *Liber emblematum* of several years before.[11] A snail slides slowly out of a cavernous lair, a place that surely is not its home. Dwelling in its spiral shell, free to go where it will, it moves at its own pace along the ground; its foot moistens and sucks as it slips forward, making its way, literally, *chemin faisant*. Antennae in the air, it ventures into the landscape it feels but clearly cannot see. The image of the common gastropod is set above and illustrates two poems in praise of discretion, the trait implied to be the crowning virtue of public duty. The first poem, a quatrain set within the architecture of the emblem, is followed by a subscriptive explanation on the folio to the right. The words turn the snail into a totem of diplomacy. The comparison that links its motto, "Secret est à louer" (Secrecy is praiseworthy), to the image is glossed in the quatrain held in the ornate frame in which are included tendrils, consoles, spirals, fleurons, and putti, as well as two stern faces in profile, each born of acanthus leaves, that stare at each other across

Figure 1. Gilles Corrozet, *Hécatomgraphie* (1541), emblem 20, "Secret est à louer," fol. D.iii v. [Typ 515.43.299] Houghton Library, Harvard University.

a set of six concentric circles enclosing an enigmatic design. Over it a winged cherub looks down as if to accord the image its benediction: "Just as the snail *[Lymas]* keeps himself in his shell, in great secrecy, so then man is held closed and concealed in discretion."

Self-containment is the virtue that would seem to be the logical correlative to the image. The poem that is set adjacent to the device on the folio to the right further explains (or, in the lexicon of theory, "unpacks") what appears to drive the analogy. Just as the little snail resides in the quiet protection of a home of its own, so then (so goes the allegory) we, too, ought in our prudence be coy and firm in our thoughts; flee evil when it threatens and make use of fortune whenever possible; come forward when danger is past; yet remain concealed, discreet, forever at ease as if at home. The snail tells us to live within our own limits and that nothing is better than to be *à tout part soy,* like Montaigne in the carapace of his château, anywhere in and about ourselves. The experienced reader of the emblem might wonder if the analogy is forced to the point where, however discreet it seems, it pushes forward in a direction other than what the gloss proposes. Snails seem more apt for a *festina lente* or a variant on Zeno's paradox than for a homily on secrecy.

That the analogy does not quite work may attest to its poetic and geographical virtue. The metaphor prompts the reading viewer to look for something else in the twisting line of the shell, a shape that seems to be an analogue of the curls and flourishes of the frame. What else is happening in the modest image? It does not take long to discern that the shell bears a resemblance to an eye, and that the parallel hatching depicting the darkness of its lair is similar to an eyeball in a socket outlined by the rocky outcropping above and the ground below. The cave and the snail are part of a zoomorphic landscape in which what is seen is what sees. The ocular snail becomes an event: a monocular shape takes form to suggest that its greater body is found elsewhere or beyond the limits of the frame.[12] Or perhaps, like a sacred being, the head and body to which the eye belongs cannot be seen or illustrated as such. The snail inches toward the space of the image of which it is a part, as the words denote, *à tout part soy.* Its two eyes, minuscule circles at the ends of its longer tentacles, would be blind eyes that *feel* the atmosphere in which they are placed; that differ from the clumps of branches atop the pruned shrub (which resembles the snail) to the left by being at once sentient, vegetable, and animal.[13] Its two eyes are reminders that a binocular view is needed to grasp the illusion of the depth of the field toward which the snail is moving. The scene shifts between the way a single eye offers a flat and haptic, almost muscular prehension of things as they are projected on a two-dimensional surface and two eyes yield a mental appreciation of a greater ambient space. The snail brings forward the two faculties, two fashions, and two ways of going about, tasting, grasping, and experiencing the world. One pertains to touch, taste, and even smell—to primitive or archaic, childlike modes of apprehension, while the other belongs to sight, a faculty we are told is of a later and higher order. In the image it would seem that the spiral, the shape that tells the viewer to turn and twist about the world, takes sight of it panoramically. The gastropod evokes the sense that the eye can touch what it sees, and in its form it brings together monocular and binocular modes of perception.

A Topographer's Lens

Without belaboring the religious ideology by which members of the Gallican church in Corrozet's milieu saw the world as a complex of hidden signs in which the *secrets of nature* bore witness to the hidden presence of God (or at least served as signs of the presence of God), the emblem can be appreciated as an intermediate form, like the snail in the hierarchy of living things, somewhere between fable and natural science; between the subject of a *blazon* and an animal whose shape inspires ocular fascination; between a subject of homily and another of scientific fantasy. It can also be seen in the tension of topography and geography. The isolated eye that the snail and its milieu both reveal and conceal has its most immediate reminder in the celebrated similitude with which Pieter Apian, following Ptolemy, inaugurated his *Cosmographia*. In his landmark manual (which saw thirty editions between 1524 and the end of the sixteenth century), Apian appends an emblem to illustrate the difference between geography and topography. In the woodcut tipped into the Parisian edition of 1551 (Figures 2 and 3), two images are contained in as many circles enclosed within a rectangular frame. Immediately below, a rectangle of identical size and proportion contains two scenes, the one to the left being almost two and a half times as large as the other. The left-hand circle in the upper rectangle encloses a map of the known world (the North Pole at the bottom and Antarctica at the top) prior to the Columbian discoveries. The three continents, viewed as they had been shown on T-O maps,[14] are in a vaguely representative view in which the river Nile dominates Africa and the Caucasus Mountains, Asia. A portrayal of the world, Apian declares in homage to Ptolemy and his followers (Johannes Werner is cited in parenthesis), is to cosmography, the description of the world and its place in the heavens, just as a portrait—seen in the circle to the immediate right—is to topography, in itself the consideration of "a few places or particular spots," without having among them any comparison with or semblance to "the environment of the earth." The city view in the lower left corner gathers some densely clustered buildings that form a defensive wall around the edge of an island on which are perched a fortress and a monastery. The expanse is seen next to two surreal shapes in the smaller frame to the right. An ear that resembles an ichnographic view of an island floats in white space next to an eye that seems to be looking at the city view to its left.

When considered together all four images share some uncommon parallels with Corrozet's image. First, the Bosch-like oculus of the *Hécatomgraphie* finds its model in the image of the organ that looks on the city view. Second, a play of monocular and binocular vision is shown in the contact of the two circles that enclose the world and the portrait of the man in profile. The double view offered by the two circles and the two *single* eyes seen in profile (which contemplate the objects before them) would be in a flat perspective. Third, the geographical images are seen, as is the emblem, as objects of contemplation.[15] The man in the upper-right circle gazes upon the world from an angle perpendicular to ours. In that configuration the sense of power emerges from the manifest difference of the viewer or viewers seen frontally as opposed to those being viewed in profile.[16] We, who look directly upon the world, also gaze upon the man portrayed who beholds the very world before our eyes. In the upper section the viewer of the world map is above or outside the world, omnipotent, like God, while the man portrayed is subject to the viewer's proximate scrutiny. One view is from afar and the other is from close at hand. Yet his angle on the world—if we see him looking at it from an angle perpendicular to our own—is similar to ours, unless of course we imagine him contemplating the globe much as a cosmographer might. Fourth, the man portrayed has striking resemblance to late Gothic images and busts of Christ, and thus brings a religious inflection to the similitude. The image is infused with theological innuendo. Christ, who would be the "pupil of our eye," offers an unobstructed and static object of our gaze while sharing an angle of view similar to ours as we look upon the world that stands in front of

montaignes, fleuues, riuieres, mers, & autres choses plus renommées, sans
auoir regard aux cercles celestes de la Sphere. Et est grandement prouffi-
table à ceulx qui desirent parfaictement sçauoir les histoires & gestes des
Princes ou autres fables: car la painture ou limitation de painture facile-
ment maine a memoire l'ordre & situation des places & lieux, & par
ainsi la consummation & fin de la Geographie est costituée au regard de
toute la rondeur de la terre, a l'exemple de ceulx qui veulent entierement
paindre la teste d'une personne auec ses proportions.

La Geographie.   La similitude d'icelle.

La Chorographie de la particuliere description d'vn lieu.

C Horographie (comme dict Vernere) est aussi appellée Topographie,
elle considere seulement aucuns lieux ou places particulieres en soy-
mesmes, sans auoir entre eulx quelque comparaison ou semblance a l'en-
uironnement de la terre. Car elle demonstre toutes les choses, & a peu pres
les moindres en iceulx lieux contenues, comme sont villes, portz de mer,
peuples, pays, cours des riuieres, & plusieurs autres choses, comme edifices,
maisons, tours, & autres choses semblables, Et la fin d'icelle s'accomplit
en faisant la similitude d'aucuns lieux particuliers, comme si vn painctre
vouloit contrefaire, vn seul oeil, ou vne oreille.

La Chorographie.   La Similitude d'icelle.

A iiij

Figure 2. Pieter Apian, similitude of portraiture and topography in his *Cosmographia* (1551), fol. 2r. Courtesy of the James Ford Bell Library, University of Minnesota.

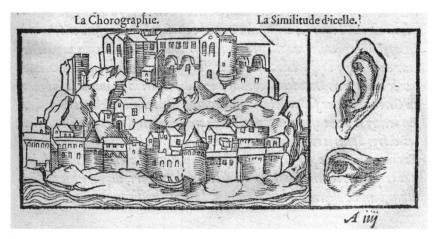

Figure 3. Detail of a city view, eye, and ear drawn from the portrait in Pieter Apian, *Cosmographia* (1551), fol. 2r. Courtesy of the James Ford Bell Library, University of Minnesota.

him. And the globe he contemplates—albeit across the border of the tondo—is total, to the point even that the intersection of the two tangents at the level of his nose assures a contact between one sacred form and another while, below, the isolated eye may be blocked from seeing what is shown to its left. It is at once an anatomical object, a detail, and an island, *like* that of the city view, a floating agglomeration of buildings. What sense can be made of things when they are beheld in detail, as various secrets of nature, becomes the question implicitly asked in this section of the illustration.

The eye of Corrozet's snail bears more than passing resemblance to the *oeil isolé,* the isolated eye of Apian's image. Whereas Apian's floats, Corrozet's seems to crawl. Each begs reflection on location and being; each asks the viewer to see how and where analogies in image and in writing can go and with what implications concerning the greater issues of situation: How do we see? With what faculties? And how are those faculties tied to matters of space and place? If being is a function of location, the commanding issue of how and what one can apprehend precedes contemplation of the greater world and the heavens. And if, as Montaigne will later imply, epistemology precedes ontology, can it not also be said that in psychoanalytic, poetic, and material senses alike, *topography precedes cosmography,* even if it is considered to be an adjunct?

The errant eye that is the snail can be construed as a topographical eye: it feels, touches, tastes (even moistens) ambient land and air. The floating eye extracted from Apian's portrait simply seems to be its own object, just as the ear above it listens to the silence of the scene in which it floats. Corrozet's and Apian's eyes inspire these questions, and they seem to be a hallmark of the topographical drive that mobilizes a great deal of creative investigation in the early modern age.

Itineraries

The task of this book is to consider the topographical impulse in the spatial and cartographic dimensions of intermediate forms, in which writing and printed images are mixed or miscible and that range from ephemeral to canonical status in sixteenth-century France. One aim is to discern and deploy a haptics of seeing and reading, indeed to retrieve the composite faculties required for the touch or taste of maps and poetry or, rather, maps *as* poetry, and vice versa. Another is to animate documents whose inherited meanings seem to overshadow or inure us to their alterity. To deploy an errant eye, not in search of things other and unknown, but to happen upon them, and to let their alterity come forward, constitutes the precondition to the *event* of discovery, in the strong philosophical sense, that is otherwise taken to be a theme or commonplace in the heritage of early modern writing. To happen upon these sites and, if possible, make events of them is what this study would like to do. Analogy, the ceaseless process of early modern reason that leads to unforeseen discoveries and relations among things, much as Apian's and Corrozet's emblems make clear, demands reflection on the nature of topography in respect to an *event* insofar as its "nexus of prehensions" brings forward the sensation of space in a process of perception in which subjectivation and objectivation are quasi-simultaneous.

The bearded man portrayed in Apian's analogy (clearly fashioned and indeed *testoné* to resemble an image of Christ) could be the very topographer whom Montaigne seeks to provide accurate descriptions of the places he or she has seen and known.[17] Topographers, like Corrozet's snail, are those who "prehend" the ambient world with their feet, eyes, ears, and other sentient organs at the very moment they give it measure with the aid of theodolites, dividers, and astrolabes. "Il nous faudroit des topographes qui nous fissent narration particuliere des endroits où ils ont esté" (203/205). [We ought to have topographers providing us with particular accounting of the places where they have been.] The pregnant remark sums up and leads in new directions many of the topo-graphical impulses felt in the richly mixed domain of poetry, painting, literature, and the practical spatial arts.[18] Montaigne will be an end point of the pages that follow, but only perhaps insofar as the work on what precedes the *Essais* will have been, it is hoped, through the "snail's eye" of the poet-topographer who writes less with the abstraction of concepts or placeless themes than with an eye, ear, and nose for detail.

In the famous eighth chapter of *Pantagruel,* the eponymous hero's fa-ther, Gargantua, who will be the title and subject of the sequel, writes to his son to encourage him to tend to the details of the world. Invoking a protoscientific attitude about gaining knowledge *(congnoissance)* of the "facts of nature," he implores,

> je veulx que te y adonne curieusement, qu'il n'y ayt mer, riviere, ny fontaine, dont tu ne congnoisse les poissons, tous les oyseaulx de l'air, tous les arbres, arbustes et fructices des foretz, toutes les herbes de la terre, tous les metaulx cachez au ventre des abysmes, les pierreries de tout Orient et midy, rien de te soit incongneu.

> [I want you to apply yourself with great care, so that no sea, river, or foun-tain remains unknown to you, and that you know every bird of the sky, all the trees, saplings and bushes of the forests, all the metals hidden in the belly of the abysses, the precious stones of the entire Orient and South, that nothing be unknown to you.][19]

The entreaty bears so much on the reader (who has just been referred to cosmographers who have accumulated narrations of the world) that much of the comic novel that follows can be read in the sense that the topographical urgency of these words becomes engaged with encoun-ter, with alterity, and with different ways of welcoming and appreci-ating its unsettling virtues. The accounts of the formative moments in the life of the young prince include prototypical descriptions that could be likened to the perception of the city view that Apian displays in his *Cosmographia.* They are set in a montage so unique that in defer-ence to what historians of cinema call the road movie, a concatenation of encounters becomes a string of events when the language depict-ing them acquires spatial presence.[20] They are especially graphic in the chapters that, burrowing into the world, become prototypical exercises in orography. They fall at the end of the novel and gather uncommon resonance when seen in view of practical manuals of topography, not the least of which is Rabelais's edition of Bartolomeo Marliani's *Topographia antiquae Romae* of April 1534.[21]

The force of Rabelais's geographical novel, although not written with or for the sake of accompanying maps and diagrams, owes much to the myriad images embodied in its writing. In *Rabelais* (1955), an inspired reading of the work that is intimately tied to his *Production of Space* (1974/1991), Henri Lefebvre remarked that for Rabelais philosophy is tributary to the creative writer's imagination. He composes with the wit and invention of images, in which space is a moving and ever-changing component of the printed word. It is through the author's distorting lens that locale is altered. Where, on the one hand, the personages encounter alterity as they go into and about the world, at the same time, on the other hand, they find themselves affected by the often-aberrant visual character of the printed words that describe their itineraries. Prototypical emblems and rebuses inhabit the printed writing. Their ideography marks the discourse that otherwise situates or even anchors the novel in the umbilical regions of the author's past, in and about the Chinonais, near his birthplace. How nascent emblems jostle and mobilize latent autobiographical plot points will be taken up in a reading of the later pages of *Pantagruel* (circa 1532 and early 1533 and its revised edition of 1542).

In his praise of Horapollo and of hieroglyphics in *Gargantua* (late 1534 or early 1535), Rabelais shares common ground with the new invention of the emblem. Felt or anticipated in the comic novel, Andreas Alciati's creation also bears a strongly topographic orientation in its French counterparts, in Gilles Corrozet's *Simulachres & histories faces de la mort* (1538) and, three years later, in his *Hécatomgraphie*. Already Pieter Apian had employed an emblematic disposition of images and texts in his *Cosmographia;* yet the French artists and designers drew the relation of the detail to the world, or the part to an unlikely unfathomable whole, through their creations of strong spatial resonance, not only as suggested by Corrozet's snail, but also in the ordering syntax, and even the difference and repetition crucial to their spatial and discursive design. The contention is that beyond their theory and practice, emblem poets and their artists make manifest a cartographic consciousness in the textual and spatial disposition of their designs. Their composite works become partial windows on a world whose extension and volume, because of the sense of detail that motivates them, are difficult to configure but invite measurement. Yet because they are read where seen, they are graphic and tabular in form.

In their collaborations the artist and writer also exploit the spatial dimensions of poetry written at the height and into the waning years of the initial popularity of the emblem. In Lyons enterprising artist-cartographers affiliated with the publishers of emblem books deploy their talents to draw city views and illustrate topographical poems aimed at retrieving antique and Italian culture and to celebrate the *genius loci* of their site of origin. The technology of the woodcut fosters such intimate correlations of image and print that the one seems to be the simultaneous and coextensive essence and origin of the other. At the same time, they also mark visual and cognitive lines of divide between the one mode and the other. The combinations prompt reflection on the relation of word and image, text and type, and figure and place. These creations range from poetry built on the theme of the conflict of country and city through the inclusion of maplike landscapes, notably in Jean de Tournes's editions of Maurice Scève's *Saulsaye* (1547) and Marguerite de Navarre's *Marguerites de la Marguerite des princesses* (1547), or in inter-mediate or embryonic atlases, especially Antoine du Pinet's *Plantz, pour-traitz et description de plusieurs villes et forteresses* (1564) before the genre was officially born with Abraham Ortelius's *Theatrum orbis terrarum* of 1570.[22] In all these works locale is crucial to the grounding vision and design. It is anchored, too, in the political and religious controversy that inspires debate over the nature of images of local places.

These transitional works, if they can be considered to be transitional objects in an equally graphic and psychoanalytical sense of the adjective, make singular use of the ocular qualities of language when it is shown and sown into the field of woodcut images. Their virtues are not lost on the Parisian poets who draw on the ferment in Lyons to launch new aesthetic programs under the patronage of the French no-bility before it splinters, in 1562, at the time of the escalation of the Wars of Religion. Joachim Du Bellay and Pierre de Ronsard, the lead-ers of the "brigade," or Pléiade, invent a new poetics of space in which topography plays a major and often changing role in the ideology of na-tional identity. In his *Olive* (1549 and 1553), Du Bellay mixes memories of Petrarch's evocations of the Rhône Valley and the Vaucluse with his own mental and physical geographies. He draws, too, on Scève's poetry of locale to substantiate the claims he makes about the excellence of vernacular French in his *Deffence et illustration de la langue françoyse* (1549). Yet it is Ronsard, in his *Amours* of 1552 and 1553, who mobilizes the sonnet in the direction of engraving and of topography in ways that French literature had never before (or would perhaps since) witness. In their performance the poems invent new and other spaces, locales that are highly autobiographical and, no less personally, cosmological. Common places are mythified to be made sensuously strange. Forests and fields teem with life through the form and force of writing. Yet the current of what might have been the "new wave" of the Pléiade from 1549 to 1555, from Du Bellay's tract and *L'olive* to Ronsard's *Amours* and their *Continuations,* changes direction and character by the end of the decade. In his *Regrets* of 1558, two years before the end of his life, Du Bellay writes of his depressive estrangement in Rome, but the melancholy is of such satirical bent that the very space of the printed page becomes an arena of creative tension. The "quadrature" or spatial syntax of the sonnets makes of each a prototypical exercise in elegiac mockery in a classical milieu. The crushing alterity represented in the collection redirects the sense of encounter that had been met in the journeys of the comic giants in the younger Rabelais.

From that vantage point the middle years of Ronsard's career, from 1558 to 1562, can be read in terms of a topographical shift, not a displacement that moves the poet from love to war, from the *Amours* to the *Discours,* but of a context in which space and subjectivity fall into crisis. In the intermediate and circumstantial pieces he writes just prior to or for the strategic placement in his first edition of his collected words— what today would be the anomaly of a young artist's self-fashioned retrospective—the poet draws poem-maps that extend their attention to the Columbian discoveries and the colonization of the New World within the local context of his own life and milieu. In a way that merits comparison with Montaigne's conflation of the indigenous other, the inhabitant of "Antarctic France," with the topographer, Ronsard looks inward and outward, always with the same ocular intensity that his earlier verse had known, to find points of juncture between known and unknown spaces. What he does in writings that metamorphose into the new genre of his own creation, the *Discours,* begins to make self-study and topography give rise to the poet's calling as a prototypical anthropologist before he yokes the imagined encounter with the "other" to issues of immediate religious conflict.

In 1585 Antoine de Bertrand set Ronsard's *Amours* to music in a handsomely designed work that mixes musical notation with historiated initials that were so ornate and carefully conceived that some of them seemed to be enigmatic emblems.[23] The letter *P* portrays a "poet" in and about the face, the foot, the arc, and the belly of the majuscule (Figure 4). The bard, seen in profile at an easel, seems to draw an image

of such contagious inspiration that it inspires him to thrust himself
aloft as he might be saying to himself—as he had in his first *Amours*
of 1553—"je me contente en mon contentement" (I take pleasure in
my pleasure). His body arches upward as his right foot pushes on a
globe. As he gazes upward he may or may not see a mottled butterfly
perched on the bar just above his eyes. The butterfly or *papillon* is a
winged counterpart to the pigeons that decorate the surround above.
All of a sudden the initial seems to be an avatar of a fantastic landscape
made from the items beginning (as they do in the title pages of the great
Larousse universel of the nineteenth century) with the same letter. The *P*
is a rhapsody of things *p*: poet, *papillon,* pigeon, picture, and portrait.
But if the portrait is that of the Ronsard remembered from the frontispiece of the *Amours* (of the poet in imitation of Petrarch, juxtaposed to
Cassandra in imitation of Laura; see Figures 35 and 36), and if it identifies him by his goatee and the laurel wreath on his balding and graying
head, it begs the viewer to wonder why Ronsard—unique, inimitable,
immortal—would be a function of an iterable graphic unit whose letter

Figure 4. Historiated *P* in Antoine de Bertrand's edition of Ronsard's verse set to music (1585).
[FC5 R6697 Gz576bc] Houghton Library, Harvard University.

qualifies him both as a *pictor* and as a *poeta*?[24] It may be that the *pied* or *patte,* the foot or paw of the letter, is the arm that unites the *P* and the *R* in what Geoffroy Tory called a ciphered or interlaced letter.[25] Like an "ambitious subtlety" of the kind Montaigne introduced in his writing, the iconic play of letter and image becomes its wit and invention. It is a signature and an incipit, a beginning of a line, and a visual summation of the work and person of its author.

Ronsard's letter has spatial virtue as much in Bertrand's edition as in the places where it reappears. One of them is at the head of the first sentence of the third volume of Abel Langelier's later printed edition of Montaigne's *Essais,* which incorporates the handwritten notes that the author had now and again inked into his copy of the edition of 1588 (Figure 5). "Personne n'est exempt de dire des fadaises: le malheur est, de les dire curieusement" (767/790). [No one is exempt from uttering banalities. The ill luck is to utter them curiously.] Most translations show that to utter them "curiously" is to "utter them too, too carefully" or with "overwrought seriousness." Can it be that the letter refers to Ronsard in the name of Montaigne? That by way of Ronsard, Montaigne is guilty of producing ornate perfidies of poetry? That we are to admire the "curious" presence of the inaugural letter, a visual *incipit* that anticipates and confirms the meaning of the statement? That in the wit of self-deprecation that belongs to every *captatio,* Montaigne is accusing himself of being too, too elegant and too ornate in the mannered style he cultivated not long after he had fallen from his horse? It may be that "curieusement" belongs to a very delicate politics of negotiation for which Montaigne had been known, between things useful and honest, as the title of the chapter indicates.[26] Or perhaps that the *Essais* belong to a poetic space that the author's printer is "unconsciously" affiliating with a poetry of prose through a fleeting allusion to Ronsard? It may simply be that in the history of the book (insofar as Montaigne died in 1592), the stratagem, if a stratagem it was, belonged to the typesetter and compositor. For the sake of this study the most likely answer is that the printed book is establishing a poetic space within the realm of its prose, and that the visual citation of Ronsard at the outset tells the reader to discern the writing in view of the deft complexities of the bard who wove together an incipient autobiography, fashioned for himself an enduring *place* in negotiating poetry and matters of state, and who left a legacy of inquiry into the world through an interminable drive to write. The topography that comes forward from the relation of the poet

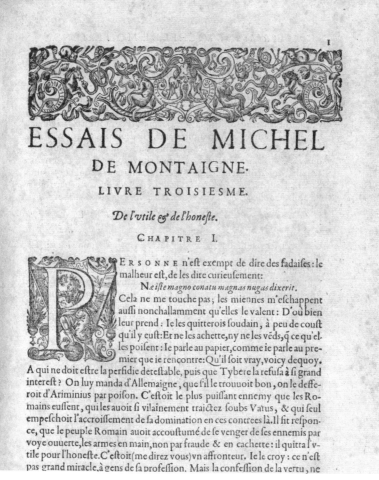

Figure 5. The first page of the 1595 edition (Paris: Michel Sonnius) of the third volume of Montaigne's *Essais*. [2004G-88F] Houghton Library, Harvard University.

and the essayist is at once literary and political. It has much to do with the places the one and the other occupy in the vicissitudes of a nation at war with itself, and it entails the allure of an invention of everyday life in times of turmoil. The essayist will defer obliquely to Ronsard in writing about the encounter of alterity, and with or through him he will reconsider things local and universal in the new and daring form of his own creation.

Here the events of self-portraiture and topography redound through the *Essais* with different inflections. How they do is, first, shown in the stress the author lays on what it means to "lodge" or "be lodged" somewhere, often in a spot that current critical idiolects call nonplaces or any-places-whatsoever, in the world at large.[27] But if it can be said that Montaigne is a topographer when he becomes any of the migrating or errant animals that make up the great bestiary of the "Apologie de Raimond Sebond" (book 2, chapter 12), he plants the human species in one of the least of all "places" in the firmament. He has the human creature (in the feminine in the French original) "logée icy, parmy la bourbe et le fient du monde, attachée à la pire, plus morte et croupie partie de l'univers, au dernier estage du logis et le plus esloigné de la voute celeste" (429/452) [lodged here, in the muck and dung of the world, attached to the worst, deadest, and most stagnant part of the universe, at the lowest floor of the dwelling and the farthest from the heavenly vault] before putting animals on higher rungs of the ladder of being. Montaigne implies that his vision of the human creature in the cesspool is taken from his own house, where, as any harried homeowner when he looks skyward, he sees swallows building their exquisite nests in springtime, in the crannies and under the eaves. Their example, like that of the spider or the fabled halcyon, two creatures of rich and noble genealogy in the literature of emblems, becomes one of many by which other perspectives are gained for us to dwell where we happen to find ourselves. The essay, it will be argued, counts among those that embody tensions of location, place, and displacement in writing that moves between memoir, natural history, and topography.

The "Apologie" begins in Montaigne's home and ends in a condition where placelessness and agoraphobia seem to be of the same measure. In its traditional reception, as a crisis of faith, as a study of fideism, and as a devastating exercise in skeptical belief, the essay finds a reassuring place in what we take to be Montaigne's world. Yet when it is seen as a piece of geographical machinery that dislocates the human being in the world that "man" has arrogated for himself, the essay serves as a guide, a *mode d'emploi,* for discovering the play of topography, cosmography, and autobiography in many of the others. In the third book the art of self-displacement is most resonant in "De la vanité" (book 3, chapter 9), the essay that distills the experience of the author's voyage to Italy in 1580 and 1581. It takes up, however, much of the emblematic material common to cosmographies, city views, and topographies when it fashions images of the writing self in movement.[28] Engaging displacement and the experience of alterity, the essay projects local space in a realm of endless duration. When Montaigne puts himself in a multilocal or siteless condition, between Rome and Paris, he becomes a topographer extraordinaire. He makes sublimely primal spaces—the matter of events—from a mixture of given places. The moment when he is born into the world in speculation about its boundaries and its duration, poetry and topography are of the same measure. The end point of this study is not where he finds himself but rather in the unsettling and intermediate area, between place and world, or the domains of the topographer and the cosmographer, where he finds that he is not. In the labyrinth or endless expanse (however the reader wishes to describe its form) of the "Apologie," the effects of poetry and topography are shrillest and most resonant.

A conceptual map, which might include a set of theoretical dividers and a measure, will be required to plot the itinerary. The study can use as a point of reference Pieter Apian's *Cosmographia,* a paradigm for theoretical and practical topography that appeals to writers, surveyors, and engineers alike. His cosmography, owing much to Martin Waldseemüller's mapping of the world and the impact of Vespucci's letters reporting his new discoveries, will serve as what later authors called a "topographical glass," less here as a "mirror in which the entire world may be reduced to graphic form" (Turner 2006, 7) than as a practical introduction to ways of treating and delineating local spaces that inform both art and practice. It will serve as a compass and as a viaticum for the study of things that on cursory glance would not be affiliated with its reflections. Insofar as it stresses the visual faculties and advises prudence and care in calculation and exactitude in measurement, it carries the ocular and ethical resolve of Corrozet's snail. The sense of topography of Apian's *Cosmographia* is embodied in Rabelais's tales of travel to local places. His *Pantagruel* will serve as a point of departure and path to Apian and beyond.

I

Rabelais: Worlds Introjected

A map of the world unfolds from a pocket book in duodecimo (Figures 6 and 7). The edition of Antonini Augusti's *Itinerarium provinciarum* by Geoffroy Tory (1480?–1533), published in 1512, is emblematic of the relation of topography to cosmography in the early years of the French Renaissance. It shows what readers familiar with Ptolemy could have made of the distinction between local and global spaces.[1] The Alexandrian's *Geographia*, translated into Latin in 1407 and known to have existed since 160 CE is often prefaced by a *mappa mundi* that stands before twenty-six projections of the countries and regions of the known globe.[2] The map, which first figures in manuscripts in the early and middle years of the fifteenth century, soon finds printed form in a variety of places. Tory's map, based on Pomponius Mela's version of Ptolemy that had been published in Venice in 1482, is tipped into the minuscule book as if to use a picture to extend the geographical reach of the text that follows. An image of the world as it had been known prior to the Columbian discoveries, the map contains masses of land and water within the broad arc of what seems to be an unfolded fan. The western border is drawn to the left of the scalloped edges of the coastlines of western Africa and Iberia. Summary renderings of Hibernia, the British Isles, and a Scandinavian Island float in the upper left corner. A vertical bar of three lines calls attention to the scheme of folds that allows the map to be included between the covers of the book. On the

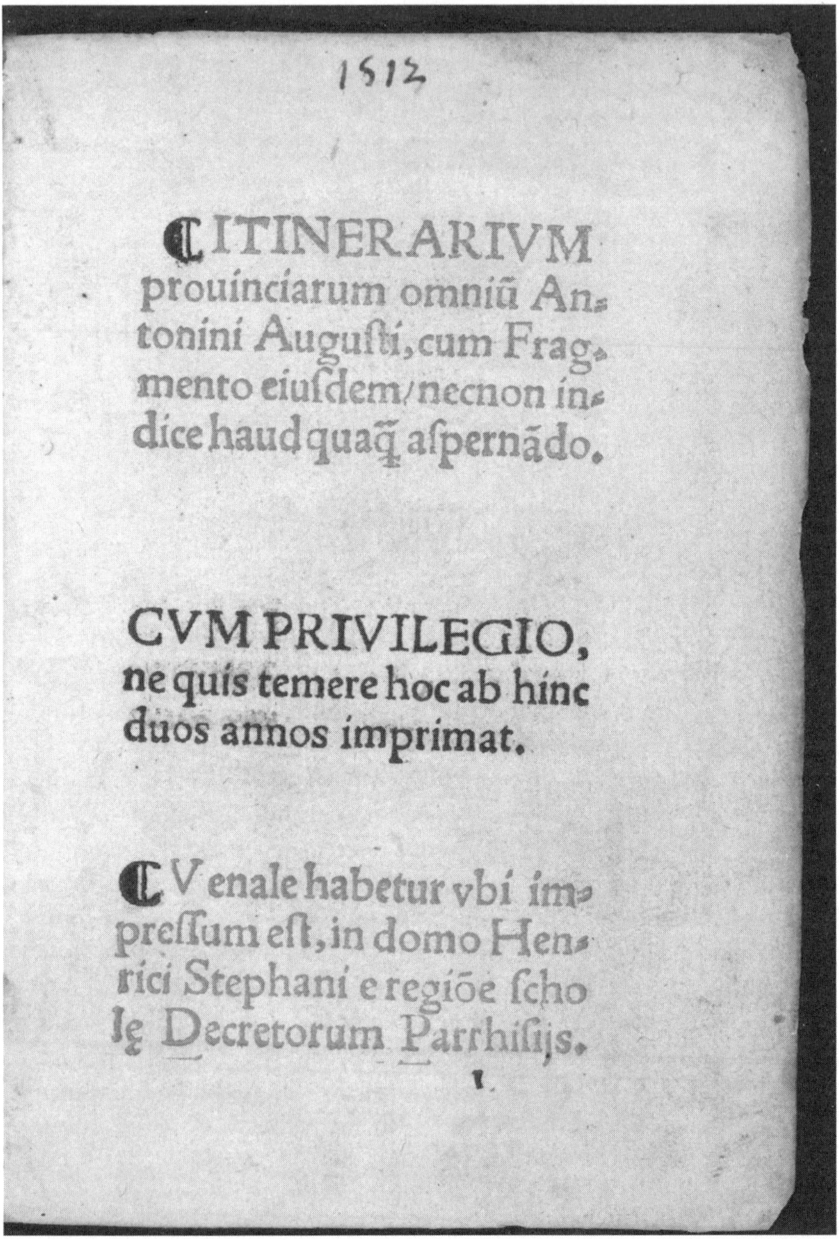

1512

℃ITINERARIVM
prouinciarum omniū An-
tonini Augusti, cum Frag-
mento eiusdem/necnon in-
dice haudquaq̃ aspernãdo.

CVM PRIVILEGIO,
ne quis temere hoc ab hinc
duos annos imprimat.

℃Venale habetur vbi im-
pressum est, in domo Hen-
rici Stephani e regiõe scho
lę Decretorum Parrhisijs.

ɪ.

Figure 6. Antonini Augusti, title page, *Itinerarium,* ed. Geoffroy Tory (Paris, 1512).
[Typ 515.12.463] Houghton Library, Harvard University.

Figure 7. Antonini Augusti, world map in *Itinerarium,* ed. Geoffroy Tory (Paris, 1512).
[Typ 515.12.463] Houghton Library, Harvard University.

right-hand side stands the Asian continent. Half of the Persian Gulf is
located on the western side, and so also is half of the Caspian Sea just
above. Most of the Indian Ocean is to the east. Within its mass the
fabled island of Taprobana (Sri Lanka) is highlighted by dense lines of
parallel hatching around its perimeters.

The lettering of the map links the eastern mass to its western counterpart. The *A* of Asia is set squarely in the Holy Land, its *S* sits in the western portion of the Persian Gulf, while on the eastern side the *I* stands
in western India, and the final *A* falls to the east of the Ganges. Africa,
whose name extends along a straight line suggestive of the equator, is

shown as a continent exceeding the southern edge of the map, suggesting that, because the context is that of the classical world, the trade routes Portuguese explorers had known and followed are not taken into account. In the upper western portion *Europa* seems to be a long peninsula drawn to include the letters that spell its name. The small continent, which dense hatch lines delineate as much as the name itself, finds vast bodies of water on three of its four sides. The Mediterranean, following Ptolemy's projection, is massively long and broad.

The bar that bisects the map (which clearly distinguishes it from the version printed in 1482) turns the mass of land and water into a function of the surrounding frame. The sixteen decorative wind heads appear so schematic and rudimentary that the traits depicting their exhalations sometimes mix with the lines of rivers (Africus blows into the southern tributaries of the Nile) or mountains (Vulturnus seems to vomit a chain of sugarloaves at the eastern reaches of Asia). They are set in six panels, the upper three in oblique perspective that, as a parergon, furnish an illusion of depth between the map proper and the quadrangular trellis-arcade in which it is suspended. The names of the winds, in Gothic script, are in accord with the point of view that each personification assumes when looking at the world in front of its eyes. Thus the austral winds that peer on the world from the south have names printed upside down and their boreal counterparts are shown right side up. Also resembling the design of the *mappa mundi* in Hartmann Schedel's *Nuremberg Chronicle* (1493), Tory's epitome map sets its itinerarium on the solid ground of an established intellectual tradition that had not yet accounted for the oceanic discoveries.[3]

The time lag between the date of the *Itinerarium* and the style of the map implies that perhaps the surrounding frame and form the book in which it is found may be as arresting as what they contain. Much like Schedel's famous map (which Noah's sons suspend or hold erect as if it were a taut bedsheet), the world is confined and of local space. In the four lower corners, in perspective, are quadrilateral plinths that support columns under Corinthian capitals. On them is placed a paneled ceiling in which beams and lines crisscross to form a grid under which two slightly different festoons extend to the left and right. Two of their tendrils extend downward to meet the wind head (Septentrio) below. The world seems contained on a classical porch or at the threshold of a garden. What the projection shows to be beyond the measure of "man" is found in a place where good company and gentility would abound.

Why would the editor put an older world map in a book about distances between local places and within an architectural decor of a room or an open terrace? The setting seems only remotely related to the classical topography the book makes available for its reader.[4] The design may be evidence of a typographer's desire to tie known and measured space to what is incalculable, and thus to approximate in a pleasing architectural composition the sublimity of a universal relation that coextends things both local and global. The map also may owe its appeal to the gridded lines of wainscoting protecting it from the rays of the sun (as evidenced by the hatching of shadow cast on the four columns). The great expanses *out there,* bearing strange and alluring names, are implied to be held within the attractively domestic spaces of terraces, loggia, and manicured gardens *over here.* The decorative festoons between the ceiling and the map seem to waft in a warm breeze. The grid of the embracing frame, which will presumably be an anchor and net for vines of ivy or summer grapes, hints, too, at the geometrical scheme, the mode of production that determines the form of the map.[5] The grounding that produces the worldly effect of the map is elegantly embodied in the architecture of a pavilion foregrounding a list of names and places of the known world.

An Encounter

Tory's map is a fitting epigraph for reflection on topography in the early work of François Rabelais. Ample and extensive studies of his geography have shown that an uncanny awareness of the changing form of the world influences the ever-changing proportions of the author's gentle giants in the worlds they inhabit.[6] Sometimes they are monstrous and beyond measure, and at other times they share the size and girth of ourselves and of our neighbors. Sometimes they inflate and distend; sometimes they shrink to a human proportion they no sooner exceed. They would fit in Tory's scheme because they belong to local places and aspire to a more animated and changing view of the world than what is given in the measure of Ptolemy. And they would be good company for those who live in the Touraine, the "garden" of France that with Rabelais is identified not only with equally local and universal strife (described in the Picrocholine war in the middle chapters of *Gargantua*) but also a utopian realm of art, literature, and poetry (at the terminus, in the Abbey of Thélème).[7] The map is fitting, too, for the play of its orthogonal lines, set in a design in which the perspective, often seen on

tiled floors, is transposed to an airy ceiling that offsets the inner sweep
of the northern and southern arcs. Meridians, set above the projection,
are shown meeting an implicit arc of latitude. The graticules that would
belong to the map itself are now seen in the ambient architecture. The
fundamental difference between the straight and curved line, an axiom
of geometry, is held within a frame in which a latent topography contains a greater whole.

Such are the effects in *Pantagruel* (1532–33) and *Gargantua* (1534–35).
For however extensive and expansive the world may be at large, its measure is found in what the eyes, fingers, hands, and feet can ascribe to
size, distance, and proportion. The bodily being of these giants, whether
inherited from popular chronicles or from narratives of antediluvian
behemoths, attests to a topographical consciousness. The protagonists,
associated with the Chinonais and its environs, mark their travels as
dutiful regionalists. Over fifty years ago, in a pathfinding chapter titled
"The World in Pantagruel's Mouth," Erich Auerbach is said to have
summed up all that can be said about Rabelaisian tensions of locality
and alterity.[8] While waiting for a torrential rain to abate in the midst
of the military campaign against the Dipsodes, the narrator, Alcofribas
Nasier, the Lucianesque chronicler who is relating the facts and deeds
of Pantagruel, cannot find shelter under the master's tongue. He forcibly climbs into his mouth. Inside, a world opens before his eyes. He
takes the first road he finds. Soon after climbing over the Dannoys
Mountains he reaches a plain where he meets a farmer who will soon
become the most celebrated cabbage planter in Western literature.

Auerbach notes that the meeting of Alcofribas and the farmer turns
topsy-turvy the convention of encounter with alterity. Eminently familiar, the man is not the frightening *other*, a denizen of antipodal
lands unknown, of Patagonian giants, who might resemble any of the
seven figures seen on the left-hand margin of Hartmann Schedel's map
(under the running head "Das ander alter der werlt") in the *Nuremberg
Chronicle*.[9] Nor is he the ilk of "cannibals" shown on recent maps of the
New World, whom, the narrator announces, Pantagruel will subdue in
a future sequel to this minuscule book. The planter hardly hails from
the limits of the known world as might "la fille du roy de l'Inde nommé
Presthan" (336) [the daughter of the king of India named Prester John],
the strange princess whom the hero, despite all promise made to the
contrary, will never marry.

Rather, the planter belongs to his *terroir*. Alcofribas's surprise is due to the encounter of a person he easily recognizes, a person with whom he can speak without encountering the slightest idiomatic obstacle. Following Auerbach, readers would correctly surmise that alterity is discovered through what it is not—through familiarity; or, failing that, that some common ground must be found *within* those who face alterity when it is ineffable: the voyager can begin, according to a psychoanalytical idiolect, to *introject* that which causes disquiet instead of suppressing its effects in what would be a counterpart, the process of incorporating—that is, tantamount to burying—the cause of anxiety elsewhere in the body.[10] When he happened upon the planter, Alcofribas was astonished. He recalls that he saluted him with gentility: "Dont tout esbahy luy demanday, 'mon amy, que fais tu icy?'" (331). [And completely astonished, I asked, "My friend, what are you doing here?"] In this first, signal or phatic moment, Alcofribas (graphically) "introjects" his initial bewilderment. He recognizes that he was afraid by accounting for his feeling of disquiet that came when he met the man; and he affirms that he responded to his own disquiet by addressing the *bon homme* in the familiar voice, as a coequal. Had he used a more formal or aggressive mode of address, the traveler would have risked causing animosity or conflict.

The farmer's answer to the narrator's query reiterates some of Pantagruel's words in the quasi-biblical encounters the hero had experienced earlier (in chapters 6 and 9). They underscore further the fantasy of introjection through the geographical form of the episode. The correlative shapes of the objects the informant mentions (vegetables that might resemble globes) establish a cosmographic context for the site and situation. "'I plant (he said) cabbages.' 'And what and how,' said I. 'Ha, Monsieur' (he said), 'not everyone can have balls heavy as a mortar shell, and we can't all be rich. That's how I live my life: and I carry them to the market to sell them in the city just over there.'" With the meeting comes the image of rows of cabbages in a tilled landscape. Like vineyards, lines of vegetables or furrowed fields, signs of culture and cultivation abound, much as they do at the edges of contemporary city views.[11] The image of dotted lines representing cabbages in a tilled landscape becomes implicitly comparable to a field of papillae, the sensitive taste buds that dot the landscape of the tongue, *la langue,* the lexical landscape on which Alcofribas is traveling, thus confirming the analogy that holds among topography, landscape, and anatomy.[12] And just as cabbages resemble testicles, so then testicles might bear resemblance

to mortar shells, and mortar shells to terrestrial and celestial globes.
Four units of comparative measure (and weight) are given in the first
words of the exchange.

And so too is the topographical distinction, par excellence, of the
country and the city. Up to this point the Pantagruelists have been
at war, far from home, where they had traveled many leagues to de-
fend the world against the Dipsodes, who had destroyed the nation
of Utopia and were laying siege to the "great city of the Amaurotes"
(327). They had set sail from Rouen, the port of departure for oceanic
passage. They had weighed anchor, too, in the presence of familiar
lands left behind, at least when an elegant lady of Paris posted a letter
to be read in transit.[13] The narrative, taking a sudden turn, follows
the itinerary Iberian travelers had followed to the Indies as they had
been recorded in Simon Grynaeus's *Novus orbis regionum ac insularum
verteribus incognitarum* (1532) that included two world maps, one by
Sebastian Münster (decorated, perhaps, by Hans Holbein), the other a
double cordiform projection by Oronce Finé (Figure 8).[14] Having gone
from known to unknown lands, from the western and eastern shores
of the African continent to imaginary lands, in most likelihood the
Pantagruelists would be far from their stomping grounds. The familiar
world discovered within the master's mouth has the effect of folding
the places noted in chapter 24 (301) into and about the landscape. The
order of Grynaeus's voyages is altered when the narrator goes on foot,
alone, and is in direct contact with the unknown, free of the protection of an armed cohort, and hence especially sensitive to the world
about him.

He is vulnerable but also better prepared to be an ethnographer. Although he might be a pilgrim who goes to the limits of Christendom—
"je y cheminoys comme l'on faict en Sophie à Constantinoble" (331) [I
walked along as they do in Sophia at Constantinople]—Alcofribas uses
analogy to judge the differences between toponyms and real places and
those of the meaning of names and their unlikely referents. A keen observer, he saw *(veiz)* great outcroppings, like the Dannoys (Danish? paradoxically flat or rolling?) "mountains" that were his master's teeth, but
he saw them in the word that he beholds in the eye of memory: "je croy
que c'estoient ses dentz" (331). [I believe they were his teeth.] The shape
and sound of the word are set against the imaginary landscape, which
in the printed account melds topography and typography. The analogy extends not only from the sugarloaves representing mountains in

Ptolemy's regional maps (of which none can be found in Denmark) to the teeth that would be their bodily correlative. The descriptive setting of "ses dentz, et de grands prez, de grandes forestz, de fortes et grosses villes, non moins grandes que Lyon ou Poitiers" (331) [his teeth, and of huge fields, great forests, of strong and squat cities, not in the least as large as Lyons or Poitiers] generates space through the repetition and variation of the fragmentary attributes of the landscape: "*dent*z . . . de *grands* prez," and then "de *grandes fore*stz, de *fortes* et gros*ses* vill*es* . . ." Comparison of the unfamiliar terrain to the geography of France is implicit in the form of the sentences that generate motion before the reader's eyes. The words present an undulating landscape in which they themselves are found and where they equally convey and are conveyed as objects and intermediaries.

The ocular character of the ethnographer's words is shown in the dialogue with the cabbage planter. He is the first person whom Alcofribas has met, or *trouvay*. That he is a "bon homme" affiliates him with the "good" or *bons* Pantagruelists and with the character of a universal Frenchman. Alcofribas is entirely—*tout*—astonished at what he encounters. In the graphic rhythm of the meeting *trouvay* carries a visual valence in its sense of "turning about and around," to *tourner tout autour*. The eye glimpses the verb and alters it by discerning in its shape not just an etymology but also a perspectival dimension. There the roving eye, like the narrator himself, discovers new horizons of form.[15] The *où*, or the "where" sought at the center of *tr-ou-vay*, is inverted in the "b*on* homme," turned right side up in "ch*ou*lx," inverted over and again in "d*on*t t*ou*t esbahy," and even "m*on* amy." The commanding question of *where* comes forward when the narrator retrieves familiarity, and where, too, the graphic traits of the chosen words articulate a space and a place in which the narrator and reader find themselves at once situated and displaced. *Où* in fact begins to redound in *choulx*, so as to be both a cabbage and a spatial marker indicating why the farmer chooses to say that his testicles (*couillons*) are lighter than a cannon shell, or why "we all can't be rich" (et ne p*ouvon*s estre t*ou*s riches).

Alcofribas's celebrated exclamation implies cognizance of the Columbian discoveries forty years after the fact.[16] The form of the remark begs the eye to pause and linger on the parenthetical marker:

Jesus (dis je) il y a icy un nouveau monde. (331)

[Jesus (I said) here there is a new world.]

The slightly profane tone in the dialogue can be taken to show that the narrator is as familiar in his words as the man who, in his second reply, suddenly swears that not everyone can have testicles as heavy as mortars. "Jesus" would put the two men on the same wavelength because, as it were, people who swear together stay together. Given the here-and-there of the locative, the statement begs us to ask where indeed *icy* might be. *Here* is *there.* What would be the "*nouveau monde,*" Auerbach notes well, is suggestive of the Touraine. If the world is turned topsy-turvy, so are the letters that convey its upheaval. Partisans of the Christian design at the basis of the book quickly note that in the parataxis, *Jesus* is the name preceding the discovery that comes at the end of the sentence. The site and the content of the exclamation suggest that the position of Christ in respect to the Columbian encounters has geographical implication: does Christendom extend to the limit of the world, to the *ecoumene,* or does it leave in its margins or beyond its purview entire populations of lost or unknown souls?[17] Or, if the graphic quality of the *lettre bâtarde* of the first editions inflects the meaning of the sentence, could the spacing of letters and their ocular play turn on the homonym "I knew" *(je sceu)*? Because of the locative parenthesis that surrounds the statement, two readings are both possible and licit.

The parenthesis that follows plays on the same issue of position. "Certes (dist-il) il n'est mie nouveau: mais l'on dist bien que hors d'icy il y a une terre neufve où ilz ont et Soleil et Lune: et tout plein de belles besoignes: mais cestuy-cy est plus ancien" (331). [Surely (he said) it's hardly new: but they all say that outside of here there's a new land where they have both Sun and Moon: and it's chock full of great doings: but this one here is older.] The humble cabbage planter seems to know his Ptolemy and Münster. In his response to Alcofribas's exclamation, beyond the continuous play of *ou, on,* and *tout,* his words teeter between those of a cosmographer and those of a topographer. The new world is completely "full" of immensely great doings over there, just as *belles besoignes* localize the places over here. The cabbage planter's unguarded assertion, "cestuy-cy est plus ancien," is based on a Socratic or Erasmian ruse—a stratagem to encourage the interlocutor to overlook the familiar terrain, one that prompts him to jump ahead to find his bearings. As soon as a greater entity is mentioned the narrator registers the landscape that the cabbage planter had indicated *icy derriere.* The farmer does not point his finger in any direction (although we imagine

him doing so). Indexical signs are felt either sotto voce or in the names
he brings to the cities in his midst.

A Meeting: An Event

Writers have often shown that to qualify as events, originary encounters
need to be reiterated: this happens when Alcofribas takes to the road
to meet the good inhabitants of Aspharage. Along the way he finds (*je
trouvay,* he says again) a companion to whom, he can, in the guise of the
tourist who is a budding ethnographer, inquire of daily life as he had in
his exchange with the cabbage planter. The initial, almost "phatic" meeting with the cabbage planter gives way to a second, with a fowler, which
rehearses what has just taken place. The bird catcher casts his net for pigeons that are implied to migrate from another world.[18] Gentle address
is used once more. "Mon amy," he asks, where do these pigeons "here"
(icy) come from? When the fowler insists that they are from "there"
(l'aultre monde), the suggestion says more than Alcofribas's conjecture
that when Pantagruel yawns the pigeons fly from the other world. Far
to the east of the Indies, and even east of Cathay, *l'aultre monde* would
be in fact America.[19] But only initially: in its iconic dimension of the
scribe's reflection that "pigeons à pleines volées entroyent dedans sa
gorge, pensans que feust un colombier" (331) [entire volleys of pigeons,
thinking that it was a dovecote or *colombier,* flew into his throat], the
dovecote refers obliquely to Christopher Columbus and his discoveries
as shown on maps and recounted in recent literature of travel.[20] In this
passage, discovery of the New World, which is shown to be yet un-
discovered, is anticipated by signs of its coming that refer to birds that
have strayed off course or have traveled from unknown places—what
French ornithologists today call *les égarés d'Amérique,* strange or different birds sighted along European flyways. In all events it is to be wondered in what direction Pantagruel faced when he yawned. The after-
noon light and prevailing winds would have the giant facing new and
untold lands. And if they are of an *aultre monde,* their alterity aligns
them with the *altérés,* the Dipsodes, seen first in the title of the book,
who are both thirsty and other, and who are now subject to the ruses,
conquest, and the evangelical ideology of Pantagruel and his band.

The image of new and expanding worlds shown within and through
the words conveying them is made clear from the beginning of the chapter. Using the formula that characterizes the peripatetics of a toponymical tale, Alcofribas recounts, "Puis *entray* en la ville" (331). [Then I went

into the city.][21] He reenacts in the first person what was said earlier of
Pantragruel in the third. The solitary entry into the city marks a strong
contrast with the quasi-royal entries accorded to the Pantagruelists
when they conquer everyone except the Almyrodes (330). The episode
now tells not of an encounter but of a formal passage across geographical obstacles and of a cultural and physical osmosis in which, in a comic
mode, the observer, like a doctor or a surveyor, calculates and measures
things in a diagnostic way. Laryngues and its twin city, Pharingues
(on the two sides of the same esophagus) suffer from a plague. "Lors
je pense et calcule, et trouve que c'estoit une puante haleine qui estoit
venue de l'estomach de Pantagruel alors qu'il mangea tant d'aillade . . ."
(332). [And so I think and calculate and find that it was the stinking
breath that had come from Pantagruel's belly when he ate so much
garlic mayonnaise . . .] The order of the action is equal to the punch
line. He thinks, calculates, and finds. *Trouve* figures a movement that
at once amplifies, pinpoints, and turns about what its subject perceives.
Were it not for the context of new discoveries the locution would bear
little of the force it shares with the whole chapter.

"Je trouve mais je ne cherche pas" (I find but I do not seek), Picasso
once said and, after him, several generations of analysts. A discovery,
he implies, results from the creative nature of an encounter that causes
it to become, in the strongest sense of the term, an event. Alcofribas
appears less to seek than to find. What he apprehends seems to apprehend him. When he goes between the great rocky slopes that are taken
to be his master's teeth, and when he climbs one of the bicuspids, he
therein finds *(là trouvay)* the finest spot in the finest of places in the
world. In them are what seems to be the site in which a garden and a
trellis, or an airy space—the surround of the world map inserted in ad-
vance of the chapter on Africa (fol. A v) in Tory's *Itinerarium*—would
welcome the visitor. The Italianate landscape in which Alcofribas lives
in bliss for four good months is countered by that of a great forest,
located below the lower lips, where brigands mug and rob him. And
when he finds himself again, in a small town where he finds work that
earns him enough to subsist, he happens—just like that—upon the
local officials (recalling Thomas More's *Utopia*) to whom he reports
his misadventures. That they remark how evil and ill-disposed are the
others "de delà" (332) [from over there] betrays a fixed idea by which
the stranger is felt to be suspect vis-à-vis the local *(deçà),* who is good.
The narrator deduces—after thinking during his errant travels in the

landscape—that, just as we have countries on this and the other side of
the mountains, so they have theirs on this and the other side of the teeth.
An ethnographic consciousness is manifest. He notes that *là,* there, he
began to think *(là commençay penser)* about the differences. Distance is
taken with respect to one's identification with a birthplace and an origin
or *terroir.* The reflection that comes with his thoughts, that half of the
world has no inkling about how the other lives—"qu'il est bien vray
ce que l'on dit, que la moytié du monde ne sçait comment l'aultre vit"
(332)—seems to be the axis on which the narration turns.

It might also be a conclusion drawn from the memory-image of
Oronce Finé's double cordiform map (Figure 8) in which the semi-
circular divide of Northern and Southern hemispheres demarcates
known from unknown worlds. The portion on the left includes Europe,
Asia, the upper half of Africa, the Columbian discoveries, and the projected extension toward Marco Polo's Cathay and the Spice Islands of
which Antonio Pigafetta had recently reported in his relation of 1526,
published in French, that recounted Magellan's first circumnavigation

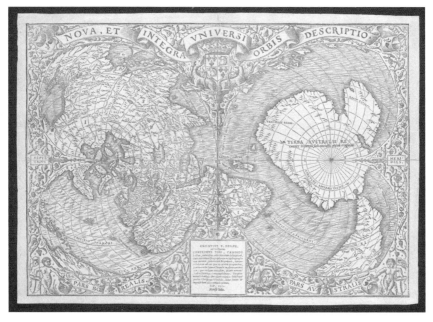

Figure 8. Oronce Finé, double cordiform map (1531), in Simon Grynaeus, *Novum orbis
regionum ac insularum veteribus incognitarum* (Paris, 1532). Courtesy of the James Ford Bell
Library, University of Minnesota.

of the globe.[22] The Southern Hemisphere to the right is filled by the presciently accurate outline of "Terra Australis recenter invienta, sed nondum plenè cognita" (Austral land recently discovered but not yet fully known), an island in which the Antarctic circle is neatly enclosed. Attached to it is what will later become Australia ("Regio patalis"), and just below a northern isthmus are the tip of Cape Horn and the Strait of Magellan. Below it stands almost all of "America," where near the bottom the land of the cannibals is noted. If the sources for Alcofribas's adventures include Folengo, Lucian, the Bible, and Galen, it may be that the visual correlation with the remark "we don't know how the other lives" can be found in the map of 1531 in Grynaeus's accounts of new voyages.

Two telling words are at the core of the episode: the world, *monde,* and *l'aultre,* the other. In few places does the writing concentrate so much on proportion, perspective, and the play of the world's familiarity and alterity. It appears that the resemblance of the mouth *(Mund)* and the globe *(monde)* that Erich Auerbach found so compelling in the episode in fact folds over the whole of the chapter and includes the narrative leading up to Alcofribas's discoveries. His solitary entry into Aspharage is a comparative analogue to what he recounts at the very beginning. A victorious expeditionary force implicitly requires the citizens of the towns it enters to praise its exploits and to welcome its presence. "Ainsy que Pantagruel, avecques toute sa bande entrerent es terres des Dipsodes, tout le monde en estoit joyeux, et incontinent se rendirent à luy, et de leur franc vouloir luy apporterent les clefz de toutes les villes où il alloit, exceptez les Almyrodes . . ." (300). [Now just as Pantagruel entered the lands of the Dipsodes with his entire band everyone was filled with joy, and suddenly surrendered to him, and with their free will brought to him the keys of all the cities where he was going, except for the Almyrodes . . ."] As Alcofribas will make clear in the sequential order of his discoveries, what happens a first time is perceived only in its repetition and variation. The event that occurs must contain or anticipate the next, or vice versa (especially to allow the text to be read as easily backward as forward). Here the billowing shifts of proportion that were witnessed in the descriptions of Pantagruel's body are found in words that apply to the distinctions of topography and cosmography.

E pluribus unum: Pantagruel is the one with many. When everyone of the camp *entrerent,* the men are in the world before they arrive *es terres.* The whole world, *tout le monde,* establishes a difference with both

terres and with *toutes les villes.* An arresting expression of *enargia* and
action, *incontinent*—a word appearing now and again in the context of
suddenness and not unrelated to the perception of space—is embedded
in the description.[23] Seen in this manner, in sudden urgency, familiar
terms are altered, estranged, *altérés,* in the sentence itself, in what is
both the ordering and the spacing of the writing. Parataxis, which a
modern philosopher witnesses as a constant reinvention of the world
(Lyotard 1986), makes Pantagruel's dialogue with Alcofribas compelling and even uncanny. The scribe has just said that upon reflecting
on the ignorance in which half of the world holds the other, he was
the first to compose a topography called the *Histoire des Gorgias.* He
happened to be the first, he avows, to write of "ce pais là auquel sont
plus de .xxv royaulmes habitez, sans les desers, et un gros bras de mer"
(332–33) [that country where there are more than twenty-five inhabited kingdoms, not counting a great arm of the sea]. The elements of a
treatment of landscapes, indeed, a topography, are mentioned.[24] If the
gros bras de mer is a great lagoon, in the text its meaning as a gap or
lacuna is not far from its similarity to the tongue or *langue* on which
Alcofribas travels. The unlikely earth that covers and protects the army
would be an empty space to be filled; the writing of the chapter—and
the *Gorgias*—is what fills it.[25]

　　The first words of the dialogue with Alcofribas seem both to ratify
and reverse the sense of the meeting with the cabbage planter. Upon
seeing him, Pantagruel suddenly asks, "Dont viens tu, Alcofrybas?" (333).
[Now Alcofribas, where are you coming from?] The same could have
been asked of the farmer and the fowler. The presence of the other
world might be felt in Alcofribas's answer to the question Pantagruel
asks him about how he sustained himself in his master's mouth. "'Et
dequoy vivois tu? que beuvoys tu?' Je responds, 'Seigneur de mesmes
vous, et des plus friands morceaux qui passoient par vostre gorge j'en
prenois le barraige'" (333). [And what did you live on? What did you
drink? I replied, "My lord, on you yourself, and I took handfuls of the
tastiest morsels that went in passing."] Anthropophagia, the defining
attribute of the cannibal, is marked before the text plays on the fact that
the narrator nibbled on food the prince had ingurgitated. "Seigneur de
mesmes vous" (333): your body nourished me, not in any ritual canni-
balism that would be related to religion, but because I needed to eat as
I might in your mouth in order to live as I did. Chronicler that I am, I
should be a faithful parasite.

In turning about and around, the letters plot the spatial fiction of the chapter. In a comparable way they play on a sense of totality that gradually disappears as the errant narrator makes his way down the mouth. *Tout le monde* rejoiced when the victors came to the land of Dipsody; afterward, signs of totality give way to evidence of fragmentary and occasional pieces of information conveyed in words related to topography. *Tout,* the adjective translating the presence of the world as a whole, in "tout le monde" (the whole world) that was joyous when Pantagruel arrives among the Dipsodes, anticipates mention of the parts comprising the sum of the world.[26] The chapter indicates the presence of the design of the T-O maps in its graphic rhetoric at the same time it causes their geography to become obsolescent in view of an expanding geography. The reminder of the schema is seen in the way the *tout* figures in this chapter and in fact bears on the material that leads up to the episode telling of the narrator's wandering.

Other Chapters, Other Realms

As of chapter 23, in which Panurge tells the famous tale about why different European nations have different scales of leagues, and in which his father, Gargantua, writes to beg him to engage in battle with the Dipsodes who have invaded the Amaurotes, the context becomes bellicose. The good giants soon go off to war. Seeing the carnage on the field of battle after they have overcome the redoubtable giant Loupgarou, the Pantagruelists discover Epistemon, their wise warrior, whose body lies inert on the ground, his head severed, much as would many of the combatants in epic tales. Panurge attends to the corpse, resetting the head in its proper place and resourcefully stitching it to the victim's neck. Epistemon comes back to life and soon tells of his visit to the nether realm where famous people of the past lead menial lives instead of suffering in flames or freezing in ice. The rich and learned listing of those whom he met becomes an occasion for festive dialogue that, in the interest of brevity, his penchant for sanctimony, or a desire to renew combat, Pantagruel cuts short. The observations about the humanity of hell give way to an illustration of meritorious treatment of prisoners.

Chapters 30 and 32 seem to be part of a staggered composition that invites comparison of the two.[27] In the latter chapter the encounter and discovery of a new world have an antecedent in the Dantesque design of the former. Both chapters are mediated by the political debate in chapter 31, in which a principal topic concerns the art of colonization and the just or unjust treatment of detainees. In fact, the first use of *colonie* in its modern sense is made here, before it becomes a topical concern in *Gargantua* and the *Tiers livre*.[28] With it, the theme of emigration, couched in the figure of the Israelites crossing the Red Sea, led by a Moses-Pantagruel to the Promised Land, carries innuendo of a white legend set in opposition to that of the *légende noire* conveyed in contemporary reports about Iberian colonization of the New World. The princely victor of French origin announces that he will lead them in the manner of a carefully organized and benevolent colony in Dipsody. Pantagruel will give to its residents *tout le pays,* "qui est beau, salubre, fructueux, et plaisant sus *tous les pays* du monde, comme plusieurs de vous sçavent qui y estes allez aultresfoys" (328) [the whole country, which is beautiful, healthy, bountiful, and more pleasant than all the countries of the world, as those of you know who have formerly been there]. The descriptions make this new world (not named as such) resemble a locale in the Touraine, that is, a setting akin to the trellis garden of Tory's world map in the *Itinerarium provinciarum*. The whole country, asserts the narrator, is more attractive than all the rest of the world.

The narrative is given to how to deal with the trophy prisoner, King Anarche. Alcofribas follows what Epistemon had reported in the preceding chapter about "comment estoient traictez les Roys et les riches de ce monde par les champs Elisées, et comment ilz gaignoient pour lors leur vie à vilz et sales mestiers" (328) [how the kings and the rich of this world were treated in the Elysian Fields, and how they then lived in doing vile and dirty work]. He turns him into a green sauce hawker and has him marry a shrewish prostitute of middle age. Pantagruel, who deals with real estate, awards Anarche a stone mortar and a little lodging *(loge)* in one of the lower streets, where he becomes, until his wife pulverizes him "like plaster," as gentle a crier of green sauce as there was ever seen in Utopia.

The chapter implements what was a state of things, an *état de lieu,* in the underworld or other world that Epistemon had visited in the short time of his death. Its attention to the wholeness of a region given to displaced souls cues the issue of encounter, discovery, and colonization in the account that follows. When Alcofribas exclaims in chapter 32, "Jesus, here there is a new world," or when he reflects that half the world has no inkling about how the other half lives, he echoes Epistemon's response to the question about whether the damned in hell are afflicted with syphilis. True, he says, the disease is abundant. More than one hundred million carry the infection. "Car croyez que ceulx qui n'ont eu la verolle en ce monde cy, l'ont en l'aultre" (324). [For you would do well to know that those don't have the pox in this world are surely afflicted with it in the other.] The formula shifts from a *vertical* order that goes with the idea of hell and purgatory to a *horizontal* and geographically informed opposition between *ce monde cy* and *l'aultre,* that is, simply "over there": Panurge underscores this idea when he locates the limitrophe area of this world not between heaven and hell but at the Rock of Gibraltar, where the Columns of Hercules are often shown as a limit, except that now the pillars are compared to bungs on the gods' barrels. Gibraltar becomes a sign of an east-west orientation, and the allusion to syphilis an implicit allusion to the Black Legend, if indeed it was known then that the Spanish conquerors were spreading the disease among the indigenous populations.

Although crude and schematic in design, the visual character of chapter 30 in the Nourry edition of Lyons is indicative of the vertical and horizontal orientations of the narrative. Printed in a cramped *lettre bâtarde,* the title acquires an emblematic aspect in its use of cul-de-lampe formatting:

> Comment Epistemon qui avoit
> la teste trâchee fust guery
> habillement par Panurge.
> et des nouvelles
> des diables, &
> damnez.

For the first time in the edition a printing device, much like the effects seen on its title page, serves as a topographic sign in consort with the descriptions that follow. The funnel-like shape of the title resembles the arrangement of the infernal world that Epistemon describes.[29] It is also related to nascent figure-poems and epigraphs (of the kind seen in chapter 27), yet here a direct allusion is made to the form of pots and jugs that are mentioned immediately below when "Pantagruel se retira au lieu des flaccons et appella Panurge, et les aultres" (321) [Pantagruel returns to the place of flagons and calls Panurge and the others]. Inherited from the *Inferno* and, more recently, Jean Lemaire de Belges's "Second epistre de l'amant vert" (in which the dead hero visits hell and comes back to tell of what he saw), the shape of the writing becomes an allegorical map for what follows.[30] In the two earliest editions (the Nourry and Juste editions of 1532 and 1533), Epistemon is announced having his head cut off, *la teste trâchee*. In the revised text of 1542 (Lyons: François Juste), whose title page is also in cul-de-lampe, a piece of visual wit takes its place. Epistemon has *la coupe testée*, that is, his "cut head off." Why the spoonerism or *contrepèterie*, and for what spatial or geographical cause? In the *lettre bâtarde, tranchée* is abbreviated by the tilde over the *a*, a pun that ties anatomy to the map of the body, *trâchee* ostensibly confused with the trachea, the passage that will have a topographical emphasis in the description of Alcofribas's travels into the lower regions of Pantagruel's gullet. It is a conduit, a road, or a passage, in the Galen-like scheme of the body. In the later variant the cut or *coupe* allows *testée* to bear myriad meaning: his cut is witnessed, *testée,* in the writing, and thus "tested" by the ocular faculties "attesting" to the truth of the event; it can even be sensed, as Béroalde de Verville would later play on the word, as a "suckled cut," a moment of originary separation rehearsed in a comic mode.[31]

The narrator had mentioned at the end of his battle with Pantagruel that Loupgarou had lost his head and that a splinter of stone had cut Epistemon's throat (320), which, it is suspected, is cause for his decapitation. Now, as in the title of the chapter, the resemblance of the alphabetical characters causes meaning to oscillate between the graphic area of the printed page and the world they are marshaled to represent. Panurge ably—*habillement*—reattaches the head with surgical expertise and finally sews it to the neck with fifteen or sixteen stitches so that it would not fall *de rechief* (322) [once again], such that the formula itself contains the head, "chief," which would otherwise once again fall to the ground. The space allotted to the description of Panurge's surgery, the application of an unguent to the stitched wound, the patient's quick return to life (Epistemon began to breathe, *then* to open his eyes, *then* yawn, *then* release a *gros pet de mesnage* [a great household fart]), is as brief (in eight short paragraphs on one page) as the narrative of his travels is ample (fifty paragraphs over four pages): meaning that in a blink a dream of other spaces is realized, and so copiously that its relation to the diurnal work—like that of a traveler daydreaming in a new world—becomes as much an ethnography as a variant on the more limited topos of the "world turned topsy-turvy."

Close inspection of Epistemon's account reveals how a world and a social space are fashioned. He describes a variety of professions that constitute an encyclopedia of everyday life in urban and rural France. The list includes, among others, cowherd, miller, salt seller, vintner, surveyor,

glassmaker, nutcracker, ropemaker, oarsman, griller, mule skinner, hay
baler, matchmaker, cooper, gondolier, tricholoma seller, weaver, chim-
ney sweep, groomer, oyster chucker, brewer, papermaker, basketmaker,
shoe shiner, ratcatcher, mole trapper, washerwoman, and ship caulker.
Some of the professions are enumerated objectively while others are not;
some are the effects of poetic frenzy—*Tarquin taquin* (Tarquin a teaser),
Nicolas pape tiers papetier (Pope Nicolas III a papermaker)—while the
sum, formatted as a list punctuated with snippets of dialogue, resembles
a gazetteer appended to a human comedy or a proto-Brueghelian picture of the world at work. Read aloud, the names of the immortals and
their vocations can be sung in the style of a Gregorian chant (much as
poet Jean Molinet constructed figure-poems that accumulate alliterative
descriptives of his patrons) or, to the contrary, the words might be cried
out in the voice of a hawker or a "sauce crier." To conclude merely that
in the thirtieth chapter the world is turned upside down would reduce
the wealth of the social geography and the poetry of the account of
life in the Elysian Fields of hell. Like the world in Pantagruel's mouth,
Epistemon's description of the lower depths is at once very familiar and
quite other.[32] The alliterative patterns of the text, if they are not taken
to belong to the network that moves between things known and unknown, make sense in the relation they establish both with Alcofribas's
voyage and the strategic role it plays in the greater tensions of theology,
topography, and geography.[33]

Chapter 34 relates how Pantagruel, having fallen ill, is cured when
he takes seventeen pills shaped in the form of copper apples. In each
are contained workers of the order of those whose menial and manual
labor was the subject of Epistemon's recent narrative of his travels to
the underworld. Armed with shovels and pickaxes, they exit from the
doors of the pills to disengage the accumulated fecal matter. Implied
is that Alcofribas had left his own stools in the giant's bowels and that
perhaps the hero's "corrupted humors" are owed to his scribe's errant
travels in the territories of his mouth. The chapter mixes humor of scale
and proportion with gentle lessons of satirical geography. Pantagruel
first takes diuretics that make him urinate so copiously that the am-
moniac salts and alkaline ingredients in his urine flow into the waters
of more than six thermal baths in France (listed as in a gazetteer) "and a
thousand other places" in Italy (of which eight are enumerated). When
the sewer workers exit from their pills, they find a great hill of consti-

pated stool that needs to be dislodged from the passage. The Nourry and Juste editions simply note that before finding the site and wielding their shovels the men looked about and around a half-league's distance to see where the "corrupted humors" were located. In the revised edition of 1534, published at the time of the Affaire des Placards and in the context, in the fourth chapter of *Gargantua* (16–17), of the description of Gargamelle's corrupted belly (she had eaten a platter of bad tripes), satire and topography coalesce:

> et ainsi cheurent plus d'une demye lieue en un goulphre horrible, puant, et infect plus que Mephistis, ny la palus Camarine, ny le punays lac de Sorbone, duquel escript Strabo. Et n'eust esté qu'ilz estoient tresbien antidotez le cueur, l'estomach, et le pot au vin (lequel on nomme la caboche) ilz feussent suffocquez et estainctz de ces vapeurs abhominables. O quel parfum, o quel vaporament, pour embrener touretz de nez à jeunes gualoyses. Après en tactonnant et fleuretant approcherent de la matiere fecale et des humeurs corrumpues. (335)

> [and thus they fell for over a half league into a horrible chasm, more infected and stinking than Mephitis (goddess of stench), nor the Camarina marshes (in Sicily), nor the fetid Lake Sorbonne of which writes Strabo: And had they not remedied their hearts, stomachs, and their wine pots (by which our noggins are named), they would have suffocated to death from these abominable vapors. Oh, what perfume, oh, what exhalation to beshit young Gauloises' facial veils. After groping and sniffing about they finally drew near to the fecal matter and the corrupted humors.]

The style of the description explodes, anticipating the force of the famous exclamation in *Gargantua* that describes the pregnant Gargamelle stuffed with rancid tripes—"Oh, the wondrous fecal matter that was swelling up in her!" (17)—just before she gives birth to Pantagruel's father.

A world expands from its own inside, both in the initial version and in the book as it enlarges with added remarks. They move from classical to contemporary landscapes and use the authority of Strabo and a play on homonymy and toponymy to turn the religiously constipated University of Paris into a gigantic cesspool. As in chapters 30 and 32 the description has the elements of an orography and thus of three-dimensional contour and relief. After Pantagruel vomits the purgative pills, the narrator takes care to compare them to globes that had been atop the tower of a local church. The play on proportion, what had been the modus vivendi of the *Cronicqs* from which much of the form of *Pantagruel* was drawn, is implemented to designate real spaces and places within a netherworld and to yoke them for the ends of satire.

An Open End

Alcofribas's discovery of a new world in his encounter and exchange with the cabbage planter has crucial bearing on the form of *Pantagruel* and the later books. The last chapter, the "Conclusion of the Present Book," falls so abruptly that it can be wondered if the author is plotting the space of future fiction, or if overtures are the matter of conclusions. Promises are made of great things to come. Alcofribas the showman, *présentateur* or *bonimenteur,* the seller of bookish elixir, opens the narrative onto the limitrophe regions of new and old worlds. As he had done in the preceding chapter, he *alters* the matter so as to make it at once reassuring—indicating where it is and how it is situated—and comically unsettling. He tells us he has heard (but not read) a beginning of his master's heroic story and now—in a second sentence—he will bring the book to a close because he is guilty, as it were, of writing while intoxicated. The remainder will be available at the Frankfurt fair, where we will *see* how Panurge was married and cuckolded in the first month of his nuptials; how Pantagruel found *(trouva)* and learned to use the philosopher's stone; how he crossed the Caspian Mountains, sailed over the Atlantic, defeated the cannibals, and conquered the Perlas (Canary) Islands; how he married the daughter of the "King of India named Presthan" (Prester John). The cannibals belong to a specific region of Brazil seen on almost every map of the New World. Their region is one where locative images are set not only in textual accounts but also on or about world maps (such as in Schedel or Grynaeus, which seem to motivate the geographical speculation of the conclusion). The mention of anthropophagia might be a concluding reminder to the narrator that he lead the next book toward the New World, even though the words are shaped as a topical *coq-à-l'âne*—a cock-and-bull, tongue-in-cheek prediction—that in all events lends substance to the mention of the *aultre monde* in the thirtieth and thirty-second chapters.

The ending and its extension say much about how the religious matter that informs the text relates to spatiality and its sentient presence. In the editions prior to 1534 the narrator bids good night, begs to be pardoned, and tells his readers not to think so much of his own faults as theirs. The tailpiece below the last words is merely signed *finis* in the Nourry and Juste editions. In 1533 and up to the edition of 1537, reference is made to the table of contents. The subsequent material in the printed text in the editions of 1534 and after extends the dialogue with the reader, propos-

ing that the matter is of gentle (or even Menippean) satire, and that it is
far more worthy of pardon than a "huge pile" of "Sarrabovites, Cagotz,
Escargotz, Hypocrites, Caffars, Frapars, Botineurs, et aultres sectes de
gens" (336–37) [Sarrabovites, Misers, Snails, Hypocrites, Cockroaches,
Thieves, Booted Brother Lechers, and other sects]. The list anticipates
that of those who will soon be refused entry into the Thelemites' abbey.
These others, who belong as much to this world as to others, include
the snail. The *escargotz* might be hypocrites because the gastropod dis-
simulates its horns (hence belongs to a regime of silence), and it might be
one of many -*gotz:* Goths, those who speak in strange tongues *(argots),*
like the "Torcoulx, badaulx, plus que n'estoient les Gotz, Ny Ostrogotz,
precurseurs des Magotz" (142) [Devout liars, tomfools, more than were
the Goths, nor the Ostrogoths, the precurseurs of the Maggoths]. They
are barbarians, if we recall, from the book of Ezekiel, Gog and Magog
(Magot).[34] A geographical line of demarcation between new and other
worlds that began to dissolve is now redrawn in theological terms, if
only for protective measure after the Affaire des Placards of October
1534. The nether regions of fire, sulfur, and brimstone are evoked where
they had just been cast as Elysian Fields; and the will to encounter,
discover, and cultivate alterity turns into flight and fear: flee and abhor
these others as I do *(fays),* he says, and you will have my faith *(foy).* It
is difficult to see whether the graphic equivocation of the words in the
earlier editions, written prior to 1534, holds in the later additions and
emendations.

In all events, in the gap between the earlier and later versions of
Pantagruel, the tensions of theology and geography could not be more
marked. Travel to new regions is revised to slant inward and to lead to
refuge in a realm of doubt. The tenor of the revised ending indicates
a shift in spatial sensibility. It may have a counterpart in the changes
brought to the beginning. The first sentences of *Pantagruel* bring forward the person of a narrator who will describe the origins and illustrious antiquity of the prince whose chronicle he is writing. In the Nourry
edition, under the title of an illustrious genealogy, Alcofribas begins,

Ce ne sera point chose [1542: chose inutile ne] oysifve, veu que sommes de sejour,
de vous remember [1542: ramentevoir] la premiere source et origine dont nous
est né le bon Pantagruel. Car je voy que tous bons hystoriographes ainsi ont
traicté leurs Chroniques, non seulement des Grecz [1542: les Arabes], des Arabes
[1542: Barbares], et Ethniques [1542: Latins], mais aussi les auteurs de la saincte

escripture [1542: Gregoys, Gentilz, qui furent buveurs eternelz], comme monseigneur saint Luc mesmement, et Saint Matthieu (217 and 1239).

[It will be neither useless nor idle, seeing that we are at leisure, to have you
remember the first font and origin whence good Pantagruel was born. For I see
that all worthy historiographers have thus treated their chronicles, not only of
the Greeks, of Arabs and Ethnics, but also the authors of Scripture, even my
lords Saint Luke and Saint Matthew.]

In the 1542 edition, "Gregoys" (Greeks) and "Gentilz" (pagans) replace
the names of the two evangelists, perhaps because the mention of historiographers of a subject greater than Pantagruel might put the author
in rivalry with authority. In the revised edition the list also changes.
"Arabes, Barbares et Latins, mais aussi Gregoys, Gentilz . . ." (Arabs,
Barbarians and Latins, but also the Greeks, Gentiles . . .) offers a different perspective on alterity. *Grecz* is lifted, perhaps shunted into *Gregoys*
to make room for *Arabes* at the head of the list. *Ethniques* is likewise
removed, possibly because the sense is felt in the *Gentilz, Barbares, et
Latins* who fill the vacancy. In 1542 those who are strange, *barbares,* find
welcome, along with Arabs and Latins, as eternal tipplers. The ecumenical gesture reaches back to include all and everyone in the historiography
as it equally carries an indication of immediacy and presence. A temporal distance is shown between the here and now of the writing of the
genealogy and its roots in remote times and places. The barbarians—the
others, *les aultres*—would be those whom the gregarious Greeks typified
as babblers because they could not understand their idiom. Arabs would
be heathens, non-Christians; the pagans, gently noted as *Gentilz,* would
resemble pre-Christian Latins. In its revised layering the first sentence
partially disavows the ethnocentric and religious bias of the new conclusion at the end of the final chapter. It also alters the genealogy to yield a
complex topography of the author's own legend and lineage: in the formula "A*r*a*b*es, Bar*b*ares, et *Latins*" is scripted the emergent signature of
Rabelais, the name that will replace its friend and its double, Alcofribas,
when the *Tiers livre* appears four years later, in 1546. However deep and
far the genealogist reaches into the gulf of time in search of his master's
origin, another name is found inscribed (or encrypted) in order, in a
highly ethnographic gesture, to embrace and make clear how all of the
others from different times, spaces, credos, and cultures are marked in the
author's own signature. No better or truer a topography of origin and of
ecumenical presence could ever be written.

As in a map both here and in the sentences that follow, the others—

pagans, heretics, Christians, infidels, believers, and agnostics—have a place on the historical map of humankind. The peoples are mixed and for that reason the world is for the better. The narrator quickly notes that after Cain killed Abel, the earth, soaked in the blood of the past, was for "a certain year so fertile in all fruits that from its loins were produced for us unique in medlars (*mesles,* the fruit of the *néflier* in modern French) that it became customarily called the year of the fat fruit *[l'année des grosses mesles]*" (217). Would their gross smell also be a sign of a productive mix of races and type? The *mesles* were, he adds, so good that everyone ate them because they were "beautiful to look at and delicious to taste" (218).

The book seeks to recover a way of encountering and cultivating others in familiar and uncommon places. Its vision seems close to the promise, seen in Tory's *Itinerarium,* shown when a world map is held within a trellis frame. The globe is given to be a place where dialogue and travel would mix anyone and everyone from anywhere and everywhere on its islands and continents. The greater world is held in similar suspension when the historiographer introjects its novelty and new expanse in the episodes telling of Epistemon's descent to the lower depths, of Alcofribas's travels and encounters in his master's mouth, and even of the workers who excavate a way through the blockage in the hero's bowels. In their accounts of discovery and encounter the later chapters of *Pantagruel* make manifest a topography of alterity. Its disquieting effects are recognized and then turned into productive dialogue and exchange. Written in view of scripture, Rabelais's narrative embraces geography, in other words, what begins from the bodily senses. That it is built on specific codes belonging to the renascent science of space will be suggested through the cosmographical writings of Pieter Apian in the next chapter.

2

The Apian Way

In the eleventh chapter of his *Cosmographie,* "Des diverses sortes de la mensure, ou especes de géographie" (Of diverse kinds of measure or sorts of geography), Pieter Apian remarks that the road pilgrims follow en route to Santiago de Compostela can easily be found on any of Ptolemy's regional maps. He situates the Galician city according to coordinates of latitude and longitude: "Et Compostelle ou Saint Jacques, auquel on faict maintz voyages, pour le corps de Sainct Iaques, contient en longueur 5. degrez 8. minutes en largeur 44. degrez 13 minutes." [And Compostela or Saint James, to which many voyages are made for the body of Saint James, contains in length 5 degrees and 8 minutes and in width 44 degrees 13 minutes.][1] Adjacent to the remark is a diagram illustrating the various standards of measure as they are used in different countries (Figure 9). A rustic woman recalling Millet's robust peasants takes a short step as a gentleman, striding eagerly, turns his head back to admire the woman while pulling her along with his right hand. He holds the handle of a rapier in his left. The sight line of the long sword suggests that he might be holding a part of his body extending to the measure of his desire. To his right a younger man, unaware of the drama behind him, poses a long staff on his left shoulder as he takes a broad step forward. He indeed might be the pilgrim who would set his feet on the path to Santiago de Compostela.

The reader learns that the three personages are demonstrative of

Figure 9. Pieter Apian, *Cosmographie* (Paris: Vivant Gaultherot, 1551). "La mesure appertenant aux pieds," fol. 15r. Courtesy of the James Ford Bell Library, University of Minnesota.

dimensio pedalis (measuring according to feet). The scene is juxtaposed to an image of twelve hands, set at various distances from one another, drawn to teach the six types of *dimensio manualis* (measuring according to hands). In the main portion of the woodcut, the characters traverse a hilly landscape in which habitations in the background stand beneath a sky whose tufts of clouds, surging out of the parallel hatching, vaguely resemble three earlobes. The young man's stride marks a *passus geometricus* (geometrical step), while that of the female peasant is merely a *gradus* or "degree." The older man in the middle makes a *passus simplex* or simple step aimed in one direction but coyly intended to go in another. The comic traits enlivening the pedagogical image are strong and clear in the Parisian editions of the *Cosmographia*. In the others (dating from

1524 up to 1550), the scene is more schematic: three bodies are shown extending their legs on flat land. Like the armless hands of the woodcut illustrating manual measure (see Figure 13), their heads are cut out of the image in order to draw attention to the rocklike shapes set between the feet of each of the characters. One stone equals one degree, two a "pas simple," and four a "pas geometricque."[2]

The Parisian edition invests two latent narratives, one of desire and the other of voyage, into its revised illustration. Following the example of Rabelais, each also underscores a topographical virtue. For unlike its counterparts from the editions printed in Antwerp, this chapter of the *Cosmographie* includes an allusion to the terminus of the greatest of all pilgrimages in medieval Europe. Published by Vivant Gaultherot, "libraire juré en l'université de Paris" [sworn bookseller at the University of Paris], its site of origin is where he lives, "demourant en la rue de Sainct Iaques, à l'enseigne P. Martin" (Latin edition: "vaeneunt apud Vivantium Gaultherot, via Iacobea: sub intersignio D. Martini"). Could it be that the insertion of the allusion to the cathedral church of Saint James in the chapter on mensuration refers implicitly, at the same time, to the site of publication of the *Cosmographie,* on the rue Saint-Jacques of the Left Bank of Paris? Or that a sense of both lived and intellectual distance is gained when the reader uses the book's origin to establish a point of triangulation (respective to wherever the reader might be) with the city and church in Galicia?

For readers of Apian's manual the answers would be affirmative. Much of the tension in the distinction that it sets forward between cosmography and topography (see Figures 2 and 3) is manifest in the relation of the body to the local and global spaces that it describes in the wake of Ptolemy, Regiomontanus, Sacro Bosco, and other canonical avatars. Here and elsewhere the *Cosmographie* concerns the measure of bodies as it does the stars and the heavens. Its style of pedagogy, made charmingly clear where it dwells on human feet, has poetic resonance: set adjacent to the image of the three pedestrians of chapter 11, the sentence describing the coordinates of Santiago reads as if it were from a hagiographic guidebook or were bearing the traces of what Michel de Certeau has famously called a spatial story.[3] The pilgrim goes to the limit of his or her body to contemplate a place interring an absent body. The itinerary is marked by a topographical register of names and places (what in mathematical terms would be entirely abstract), acquiring a bodily aspect when the voyage is plotted according to the way in which hands or feet touch and extend themselves on the surface of the earth. This tension of mathematics and tactility marks much of the *Cosmographia,* and, it will be argued in this chapter, it makes the work a fertile manual of reference for poetic topographers of the greater part of the sixteenth century.

A Book and Its Fortunes

Fresh and vigorous, the context from which the manual emerges belongs to mathematicians and cartographers interpreting information coming from travelers having visited the Orient and the New World. It is one in which the cosmographer creatively imagines and recreates the voyages from shards of information mixed with inherited interpretations of world space. In this way it has been remarked (Buisseret 2003, xi and 15) that with the impact of cartographical representations in the sixteenth century, a new and unforeseen sense of region and locale is born. As soon as printed cosmographies begin to circulate, a topographical sensibility, a feeling of bodily attachment to a specific place, marks early modern consciousness. The observation can be extended to the emerging literary world in which the growth of the "author," a personal signature betraying a sense of appurtenance to given cultural and linguistic spheres—to communities that he or she invents through measure and imagination—takes place in synchrony with new spatial representations of the known and unknown world.[4] The emergence is especially made manifest in Apian. The *Cosmographia* was published in thirty editions in four languages from 1529 to 1609 (Ortroy 1901/1963). Based on Martin Waldseemüller's *Cosmographiae introductio* (1507; Ronsin 1991), which included Amerigo Vespucci's four letters concerning the New World, Apian's work remains an eminently accessible illustrated manual, a piece of footwork for pedagogy.[5]

It invites contemplation about where its readers might find themselves in the world. Its form, design, and evolution betray what might be called its topographical impulse. The *Cosmographia* ultimately begs its readers, presumably students in geography but also curious readers of images, to take cognizance of the world by asking where their bodies (or any of its parts) are located in respect to other places. The book implicitly asks from where or what point of view the reader is learning the implications of "location" with respect to the greater world.[6] Further, as a book of its time it varies from edition to edition, not only in the placement of the text and its images, but also in the material that

Apian's gifted student, Gemma Frisius, adds after 1530. After 1544, for
explanatory purposes, Frisius folds into the book a truncated cordiform
world map to lay thematic stress on wind and terrestrial movement.
Topographic material concerning triangulation figures in addenda that
enhance its locative orientation. Rudimentary images of men surveying
the landscapes in which they find themselves seem to anticipate many
that are found in other treatises, from Oronce Finé and Jacques Focard
to Albert Foullon.[7]

The reader gains the impression that surveyors are looking about their
milieus and taking measured note of the rich expanses before them. If
a single innovation emerges from the initially Ptolemaic tenor of the
work, it is found in the primacy accorded to the sensate experience of the
"measure" of local space as it had been noted in the sentence concerning Santiago at the rue Saint-Jacques. The same is gained through the
narrative and figural order of the book itself. Apian's reader is prone to
see (and to find) a geographical identity that is paradoxically fragmentary, local, and—with an uncommonly attenuated presence of religious
material—universal. The means and effects of its construction of topography become an emerging agenda in the evolution and variation that
mark the *Cosmographia*. Woodcut illustrations, volvelles, and maps in
the text become cause for wonder and a sense of displacement.

Following the illustrated title page, the book uses its combinations of
text and woodcut images to move to and fro from universal space to locale and *terroir*. The contents signal a movement from astrology to cosmography that runs from Apian's celestial domain to Gemma Frisius's
description of "regions and countries by geographical artifice" and a
section "on the use of the Astronomical ring," by the same Gemma, "le
tout avec figures à ce convenantes, pour donner intelligence" (the whole
with fitting figures to bring intelligence) to the design.[8] In contrast to
the first French edition of 1544, in which a rondeau and a fourteen-line
poem are dedicated "to the honor of the book" and crafted to make a
sales pitch,[9] the Parisian editor and his team aim at a higher, contemplative sense of the world:

> Au Bening Lecteur
> Amy lecteur, Apian veult ensuyvre
> Le seigneur Dieu, declairant ses haultz faictz,
> Car il *descript & pourtraict* en ce livre
> Les cieulx, la terre, & les orbes parfaictz;

D'astres luysantz, recite les effectz,
Les mouvementz, puissances de nature,
Et monstre a l'oeil, par reigle & par figure,
Les regions de la terre habitable:
A celle fin qu'humaine creature
Contemple en soy ce qui est profitable. (fol. ai v; emphasis added)

[To the Worthy Reader
My friend dear reader, Apian follows
The path of our Lord, declaring his worthy deeds,
For he describes and portrays in this book
The perfect heavens, earth and planets,
The movements, powers of nature,
And he shows to the naked eye,
In diagrams and figures,
The regions of the habitable earth:
To whose end the human creature
Contemplates in itself what brings profit.]

The eight-line piece that follows, by Fran[çois] Barat, "native of Argenton in the Berry," juxtaposes its author's site of origin to the myth of Endymion and a sense of our need to admire the globe:

D'Endymion fut la Lune amoureuse
Iadis, Lecteur, par grand'affection:
Car luy premier, sa nature fameuse
Rendue auroit par noble invention.
     Sera ce donc cas d'admiration.
Si maintenant la Terre est affectée
A Appian, qui l'ha sans fiction
Tant vivement paincte, traictée, & notée.
                    Grace & Labeur (fol. ai v)

[With Endymion was the Moon in love,
Once, Reader, with great affection:
For he first, in his famous way
Would have portrayed it with noble invention.
     So thus here is a work of admiration.
And now the world is assigned
To Apian, who, without fiction,
So vividly has it figured, drawn, and noted.
                    Grace and Labor.]

In both poems the art of description, what will be taken in a topographic sense, is contrasted to the survey of the matrix or sphere in which local places will be found. The gist of these words follows that of the invocative song (in the Latin editions of both Antwerp and Paris), in which the reader is asked in a litany to behold the thousand places, cities, and towns that are contained in the book under the heavens.[10] The preface addressed to the reader, situated between the title page and dedicatory poems, makes much of the difference it establishes between cosmic and local spaces. In the Parisian edition, words about the origins of astronomy go to and from practical application in other arts and sciences, notably for

aucuns qui suyvent les choses polyticques & civiles, s'appliquant totallement a *bien entendre les loix des villes & Regions,* & celles qui concernent le faict de guerres, les aultres s'addonnant à lire la saincte escripture, selon l'intelligence de Saincte Eglise, affin que par icelle [people] soient incitez & induitz de plus vertueusement vivre. Et s'il y en a qui *aultrement* le facent, devons prier Dieu les vouloir tellement amander, qu'ilz ne prennent en vain la saincte parolle. *Aulcuns* commes Rhetoriciens, & telz, semblables ingenieux, & subtilz esperitz, voluntiers par grand affection lisent histoires, & poesies. Ausquels ce present livre de Cosmographie d'Apian (traduict de Latin en François) sera tres expedient & necessaire, veu que *la Cosmographie est la droicte sente & chemin,* a la science d'Astronomie. Car on n'y *trouve* les cours des cieulx, & planettes: la *situation* des elementz, la haulteur du Soleil, l'augmentation & diminution de la Lune. Encores y est contenu la forme et maniere comment on peult faire la description de *toutes regions ou places que nous voulons. Aussy y trouvera on la vraye situation des quatres principalles parties du monde.* (fol. 2r–v; emphasis added)

[some who follow political and civil matters, dedicating themselves entirely to understanding the laws of cities and regions, and those concerning the art of war, and others committing themselves to reading scripture according to the intelligence of the Sacred Church so that people will thus be induced and encouraged to live more virtuously. And if others do otherwise, we must pray to God that they amend themselves in order not to take his word in vain. Others, rhetoricians and such, ingenious and subtle minds, eagerly read narratives and poetry: to whom Apian's present book of cosmography (translated from Latin into French) will be very expedient and necessary, seeing that cosmography is the straight and direct road and path to the science of astronomy. In it are found the course of the heavens and planets; the situation of the elements, the height of the sun, the growth and waning of the moon. And even contained within is the form and manner by which description can be made of all the places we wish. Thus within will be found the true situation of the four principal parts of the world.]

A pitch that a savvy editor directs to a general public, the preface indicates more than it states. Implied is that geometry of the heavenly bodies would be a model to redress imperfection and guide local policy. The reader of the cosmography will forcibly become a topographer or chorographer with a firm grasp of local knowledge. He or she will have

God's machine in mind and will thus impart its beauty to "others," suggested here to be denizens of worlds outside Christendom, who might be found to the east and west of the places shown in the book—and all the more insofar as it stakes claim to knowledge of the New World. The paratext suggests, further, that those who describe the ambient world in prose or verse—they would be topographers—will benefit by arching their words skyward, along the *droit chemin* that will later become the Cartesian path to self-edification. But no sooner does "the science of astronomy" seem to be a goal than mention of the "four principal parts" of the world brings the reader back to earth.

Site, situation, and alterity redound. The printer's arrangement of the final sentences in the design of a cul-de-lampe further *locates* them in the place of the book, at an ostensive vanishing point of its origin in Paris. Its reassuring familiarity, shown in print on the title page, is offset by the allure of the greater world's marvel and alterity. After noting, by way of affiliation with Vespucci and Waldseemüller, the three continents and the island of America (the latter containing Peru, having been found abundant with gold), the text lays further stress on its topographical character:

> Et en oultre sont escriptes les villes & places plus renommees & principalles, qui sont situées esdictes quatre parties du monde. Encores beaucoup d'estranges choses, coustumes, & monstres prodigieuz, tant d'hommes que de bestes, qui illec ont esté trouvez, & veuz, comme en la deuxieme partie de ce present livre plus amplement sera declairé. D'avantage aussi sera il au Pylotes, maryniers, & mesureurs de terre moult ydoine, expedient & necessaire [aux] amateurs de Cosmographie estantz a l'hostel en leur estude, pourront perlustrer & voyager seurement tout ce circuit de la terre, sans grand coust, n'aucun dangier, ce que aultrement ne se peult faire n'accomplir sans grand despence & peril de sa vie. Ausquelz je soubhaitte toute prosperité & salut, par la bonté & grace du createur des cieulx & de la terre.

> ***

> De Paris, lan de
> Nostre Salut,
> 1553   (fol. 2v)[11]

[Furthermore the most well known cities and principal places are described, that are situated in the four said parts of the world. And even many strange things, customs, and prodigious monsters, of both man and beast that have been found within, and seen, as it will be amply declared in the second part of this book. And too, it will be useful for pilots, mariners, and surveyors of land, and expedient for amateurs of cosmography who stay in the confines of their study, who will be able to go about with assurance and travel all around the world, at little cost and without danger, what otherwise can be neither accomplished nor done without great expense and peril of life. To whom I wish all prosperity and salvation, by the goodness and grace of the creator of the heavens and of the earth. From Paris, in the year of our salvation, 1553.]

Both the Latin and the French texts of Gaultherot's editions fittingly address the disposition of the text and woodcut image on the title page (Figure 10). The words of the title, also set in a cul-de-lampe disposition, move from the greater spheres of cosmography (in bold uppercase letters) to the names of Apian and Frisius (noted to be from Louvain) and mention of the treatment of all the places in the world. The words funnel toward the upper pole of an instrument Apian later calls (fol. 22v) an azimuth, above which in the Latin edition Frisius's name appears. The title descends to the upper pole of the image from which lines of latitude radiate to meet the degrees of an equator that separates the semicircle they draw from an ornate celestial hemisphere designating the constellations of the heavens on a black background. A left hand emerges from the lower edge of the frame enclosing the illustration. It holds a handle to show that it is an instrument. The image suggests that the body to which the hand would be attached is absent but entirely celestial: yet, graphically, it is "A Paris," the site noted just below. Would the hand belong to the editor from Paris or to a deity who overlooks the heavens? That it can be both "here," in and about the book, and "there," over and above the empyrean, illustrates a tension of topography and cosmography that will run through the book.[12]

Topography and the Body

The isolated hand of the image is not unrelated to the first and most dramatic and telling of all of the chapters in the *Cosmographia*. "Quid sit cosmographia, et quo differat à Geographia & Chorographia" [what is cosmography, and how it differs from geography and chorography], the distinction with which Ptolemy inaugurates his *Geographia,* is repeated here, but with the new difference that three woodcut images further enable its contemplation. In the Parisian editions two woodcuts

La Cosmographie de

PIERRE APIAN, DOCTEVR ET

MATHEMATICIEN TRES EXCELLENT,
traictant de toutes les Regions, Pais, Villes, & Citez du mon-
de, Par artifice Astronomique, nouuellement traduicte de La-
tin en François par Gemma Frisius, Docteur en Mede-
cine, & Mathematicien de l'uniuersité de Louuain,
de nouueau augmentée, oultre les precedétes impres-
sions, comme l'on pourra veoir en la page suy-
uante. Le tout auec figures a ce conuenā-
tes, pour donner plus facille
intelligence.

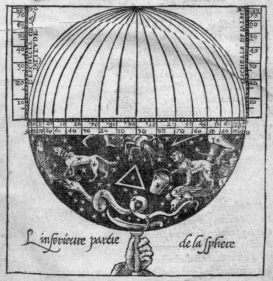

A PARIS.

Par Viuant Gaultherot, libraire iuré en l'uniuersité de Paris, demou-
rant en la rue Sainct Iaques, a l'enseigne S. Martin.

1551.

Figure 10. Pieter Apian, *Cosmographie* (Paris: Vivant Gaultherot, 1551), title page. Courtesy of the James Ford Bell Library, University of Minnesota.

are set adjacent to each other on facing folios. On the left (fol. iv), the geography of the earth and heavens is shown above (Latin edition) or below (French edition) an isolated eye out of which radiate lines that draw across and contain an earthly sphere, half-illuminated, and half in hatched lines of shadow. The lines reach the diameter of a greater heavenly counterpart, illuminated and shaded in the same way (Figure 11). Beyond being led to wonder whether what the eye beholds is based on intra- or extramissive vision, the spectator notes the presence of the perfect circle of the eye's pupil whence the eight lines emerge or converge. In the Latin edition, the eye floats over "Geographia quid," the emblematic superscription over the explanation of the distinction between cosmography and geography. In the French it stares below the running head, "la première partie," suggesting that the eye can be a first faculty. Geography deals with the "principal and known parts of the world" that comprise its entirety, which the earlier French edition (Antwerp, 1544) shades with poetic innuendo: "Geographie (come Vernerus en sa paraphrase dict) est des principaulx lieux & parties cogneues de la terre, d'autant que d'icelle tout lentier [sic; 1553: l'entier] monde est constitué, comme des choses plus renommées, qui a icelles parties de la terre sont adherentes, une description ou paincture & imitation" (f. iii v). [Geography (as Werner states in his paraphrase) is of the principal and known parts of the world inasmuch as from these the whole sum of the world is constituted, as well as the most renowned things, to which these parts of the earth adhere: a description or portrayal and imitation.] Here the deictic icelle echoes ciel, the sky above, and lentier the lenticular aspect of the image shown in the figure of the eye. The tonic force of "description ou paincture & imitation" at its end bends the sentence in the direction of topography.[13]

The French continues to move to and fro in dialogue with the image. In the edition of 1553 the eye contemplates what seems to be a pure abstraction, the earthly sphere that could be a continental mass but also a cloudlike and scarcely identifiable space. The carefully drawn pupil of the discerning eye would be a microcosm in which the two detached and superimposed spheres are reflected. The "description ou painture [depiction] and imitation" that would be the matter and substance of the woodcut remains bathed in abstraction. By strong contrast, the Antwerp editions (in French and Latin), while more crudely drawn, respond more directly to the text. A more approximate eye beholds a world on which a landscape can be seen and on whose surface, top and

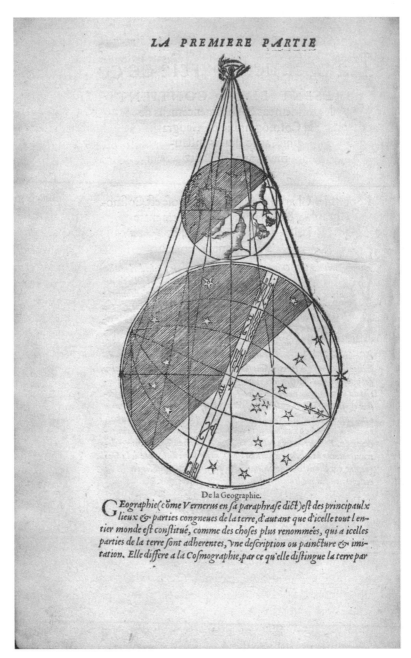

De la Geographie.

G*Eographie(côme Verneus en sa paraphrase dict)est des principaulx*
 lieux & parties congneues de la terre, d'autant que d'icelle tout l'en-
tier monde est constitué, comme des choses plus renommées, qui a icelles
parties de la terre sont adherentes, vne description ou painčture & imi-
tation. Elle differe a la Cosmographie, par ce qu'elle distingue la terre par

Figure 11. Pieter Apian, *Cosmographie* (Paris: Vivant Gaulherot, 1551). "La premiere partie, de la geographie," fol. 1v. Courtesy of the James Ford Bell Library, University of Minnesota.

bottom, two human figures walk. Adjacent to the Ptolemaic image is a world map in the form of a globe on which are marked and drawn the continents of Europe, Africa, Asia, and America (the name placed on the southern continent). In all editions the text clearly underlines that it is up-to-date by having the image designate the world as it is known. In 1544 the fact is further confirmed by the next sentence: "Et est differente a la Cosmographie, car elle distingue la terre, & determine, par montaignes, fleuves, rivieres, mers, & aultres choses plus renomees, sans avoir regard aux Circles celestes de la Sphere" (fol. iii v) [and is different from cosmography because it distinguishes the earth and determines, by mountains, large rivers, rivers, seas, and other more renowned things, without considering the celestial circles of the sphere], details the woodcut displays in the line of the Nile in Africa and the tufts suggesting the presence of the Caspian Mountains in Asia.

The sentence that follows, of capital interest for poets and artists seeking to produce "worlds" with which they rival the earth and heavens, implies that *observation,* what the monocular eye performs in the image above (in the Antwerp editions) or to the left (in the Parisian counterparts), whether historical or aesthetic, stands at the beginning of any description:

Et est [la Geographie prouffitable] grandement prouffitable [le plus a] à ceulx qui desirent parfaictement sçavoir les histoires & gestes des Princes & aultres fables: [& fictions aux histoires samblables. Car]: car la paincture ou limitation de paincture facilement maine a memoire l'ordre & situation des places & lieux. (fol. iii v)

[In Latin as] Iisque maximè prodest, qui a[da]mussim rerum gestarum & fabularum peritiam habere desyderant. Pictura enim sev picturae imitatio, ordinem sitúmque locorum ad memoriam facillimè ducit. (fol. iv, fol. 2r. Paris)

[And geography is most advantageous for those who want to know perfectly the deeds and stories of the princes and other fables and fiction in similar histories. For painting or the imitation of painting easily lends to memory the order and situation of places and sites.]

In today's terms Apian is noting that "locational imaging" conduces to narrative. But where there is narrative, space must be identical to or congruent with an active agent, something or someone (as in Aristotle's *Poetics*) that moves from place to place. For this reason, even though the text is close to its source in Ptolemy and Johannes Werner, the human figure intervenes, not only in the gist of the citation, but also, indeed crucially for the observant and wandering eye, is visibly shown in the strategy of the woodcut illustration. In the Parisian edition the sentence continues:

> & par ainsi [Et par ainsi] la consummation & fin de la Geographie est constituée au [en le] regard de toute la rondeur de la terre, a l'exemple de ceulx qui veullent entierement paindre la teste d'une personne avec [avecque] ses proportions.

> Consummatio itaque & finis Geographiae, totius orbis terrarum constat intuitu, illorum imitatione, qui integram capitis similitudinem idoneis picturis effingunt.

> [Thus the summary view and end of geography is constituted in the gaze upon the earth in its entire rotundity, in the example of those who wish to depict the head of a person in its proportions.]

Meaningful images and narratives are made by spatial measure, by a *gaze* cast upon the "entire rotundity" of the earth. The gaze on the globe must be directed at the same time upon the head of a person, who could indeed be he or she who gazes. The text suggests that wholes and parts must be discerned at once and that a grasp of the heavenly spheres can be contemplated by imagining an artist or a writer "describing" a person's head. An abyssal condition is suggested, too, if the isolated eye at the origin of the image to the left is inferred to belong to that of the man who is portrayed on the right.

The famous image that follows the text both extends and mediates the tensions initially evinced in the words. Where the image on the left distinguishes geography from cosmography, the two woodcuts on the right are indicated to delineate geography from chorography. In the editions from Antwerp a fairly crude spherical world map (north pole at the bottom and the south at the top) is placed in a square. Visible but unnamed are the continents of Europe, Africa, and western Asia (Taprobana floats in the sea to the north and east of a tip of Antarctica). In the square to the right a pensive man, his forehead balding and his chin graced with a goatee, is shown in three-quarter perspective as if, lost in thought, he looks both inward and outward. Elaborate parallel- and cross-hatching lends amplitude and depth to the face. The French text below implies that the man might indeed *personify* geography at the same time he is an object of its comparison to painting or description. And, as the text notes, he both contemplates *(consydere)* and gazes *(regarde)*:

> Chorographie (comme dict Vernere) [laquelle] aussi est appellee Topographie, elle considere [consydere ou regarde] seulement aucuns lieux ou places particulieres [particuliers] en soy-mesmes, sans avoir entre eulx quelque comparison ou semblance [samblance] à [avecq] lenvironnement de la terre. (fol. 4r)

The Latin is more succinct:

Chorographia autem (Vernero dicente) quae & Topographia dicitur, partialia quaedam loca seorsum & absolute considerat, absque eorum ad seinuicem, & ad universum telluris ambitum comparatione. (fol. 2r)

[Chorography, as Werner states, is also called topography because it considers only some places or particular sites in themselves, without there being any comparison of semblance between them with the environing earth.]

In the Antwerp editions the man's binocular view seems well fitted to discern the universal "ambit" or circle of the worldly sphere as well as the parts its image displays—including sugarloaves that could mark the Caspian Mountains, the Himalayas, the Alps, the Pyrenees, and the peaks at the source of the Nile (and even tufts of trees along the North African coast).

In the Parisian editions the redrawn image carries different inflections. The man portrayed, shown in profile, bears resemblance to contemporary images of Christ. He is shown gazing directly at the world sphere in front of him, which in this instance is a carefully rendered woodcut in which the three continents are named and where "La Mer Oceane," of vaster proportion, is pocked by the islands of Taprobana and Madagascar. The two circles in which the world and the man are enclosed establish a binocular presence in the frame. The Christological—both sanctified and sanctifying—figure seems to squint at the world he beholds, thus focusing on what he can be assumed to be both seeing and reading.[14] Yet, because of the double foci of the two circles, the reader's gaze seems to "consider and see" what, from his perspective, the Christ-like sitter could not. If that likeness of the man is as it seems, he would be the "pupil of the eye": but also the personification of the sacred quality of the world as it is both known and unknown. A sense of aura is invested in the comparison of geography (the map of the world) to what contemplates the latter and at the same time both embodies and discerns topography.

The text that completes the distinction between geography and chorography underlines further a tension that causes the two categories to blur. For *(Car)* Topography (uppercase in all editions, thus personified and typified) "demonstre toutes les choses & a peu pres les moindres en iceulx lieux contenues, comme sont villes, portz de mer, peuples, pays, cours des rivieres, & plusieurs autres choses [samblables], comme edifices, maisons, tours & aultres choses semblables" (fol. iv r) [demonstrates all things and generally the slighter ones contained in these places such as are cities, seaports, populations, countries, the course of rivers and several other things, namely, buildings, houses, towers and other similar things].[15] The demonstrative or deictic formula *(iceulx)* indicating where a place might be—here and not there—contains the places *(lieux)* themselves. Similar *(semblables)* things seem to proliferate by virtue of the similitude. Contemplation (what goes with the figure of Ptolemy who is usually shown looking skyward) is countered by the horizontal gaze that is suggested to be selective, which is exactly the nature of the operation that informs the juxtaposition of a city view to an isolated eye and ear. In a surreal way the reader (and perhaps the sitter above, regardless of whether he is of secular or sacred aspect) is asked to make an analogy between two types of detail. In the Antwerp editions a wall and fortified town are perched on a hillock over a plain below (a tuft of grass is visible adjacent to the lower-left corner of the frame). The area of the rectangle in which it is situated is the same as that in which the broadly hatched figures of the eye and ear float to the right. The words that introduce the image confound the site of the viewer as well as the organs said to perceive the city view: Thus the aim of topography "s'accomplit [sera accomplie] en faisant la similitude d'aucuns lieux particuliers, comme si un painctre vouldroict] contrefaire ung seul oeil [oyel], ou une oreille" (fol. iv r) [is accomplished by making the comparison of a few particular places, much as a painter would depict a single eye or an ear].[16] In the drift between word and image the eye and the ear—*oyel* and *oreille,* the one being graphically in and of the other—can be in themselves "particular places," thus perceiving or perceptive sites, and all the more in the blurring of desire and rectitude (if *droit* is seen inhabiting *vouldroict*). Areas that are the objects of sight and audition also have the capacity to see and hear. At once inanimate and animate, they are, in the philosopher's description of the consciousness vital to the sensation of an event a "nexus of prehensions."

In the Paris edition the very finely rendered ear and eye are compared to the view of an island-city that floats on a mass of waves (see Figures 2 and 3). Its lower surround, a series of walls topped by crenellated cornices and punctuated with towers, gives way to a basilica or monastery replete with a barbizon, and is indeed of a mass reminiscent of a high place, much like the papal palace at Avignon. The curvature of the parallel hatching marking the hilly landscape causes the scene to undulate in rhythm with the waters below. The character of an island or of utopian isolation is so manifest that it marks the eye and the ear to the right: an aural and a visual island become in themselves oneiric landscapes in which the patient—and contemplative—viewer can get lost. The distinction that might have been clearly drawn in Ptolemy, and that would be made bolder by virtue of an allegorical or emblematic illustration, is creatively confused.

The detached ear and eye, especially those of the Paris edition, float indeterminately where they are said to be organs of locative perception. They might belong to the recent tradition of the bodily blazon in which poets of the same moment select an organ in order to fashion a paradoxical or enigmatic encomium of its virtues (Schmidt 1953, 293–364). Or they can be seen related to the art of anatomy and therefore of autopsy. They seem to gaze upon and listen to the matter that "describes" them as they are. In their strange detachment the eye and the ear bring to the greater work a fantasy—that could be delightful as frightening—of bodily dismemberment taken to be integral to topography itself. Elsewhere in both the *Cosmographia* and Apian's *Cosmographia introductio* (1532 and later editions), many of the distinguishing agents in the illustrated matter happen to be detached hands, arms, or legs that appear saturated with theological residue. The body and mind that animate them are nowhere to be found. Although they are entirely topical and conventional, the presence of the inaugural similitude causes the reader to wonder whether the presence of the body in contingent or local space is implied to be opposed to its absence in the cosmographic arena.

How a bodily part enables the "prehension" of tactile perception of topography is shown in the maps inserted at the beginning and the end of the eighteenth chapter of the *Cosmographia*. In every edition a map of Greece and its ambient islands is inserted to illustrate the topic of chapter 17, "Quo differunt insula, peninsula, isthmus, & continens" [the distinctions of island, peninsula, isthmus and continent]. The French edition of Antwerp (1544) lists England, Sicily, Rhodes, and Java as demonstrable islands (fol. 27v), while those of 1550 and onward, published in both Antwerp and Paris, replace *Engleterre,* the first of the series, with America. In 1544 the *terra firma* synonymous with a continent includes Hungary, France, Spain, and Germany, while later editions list Misnia, Pannonia, Boiaria, and Saxonia. In every edition the map is only indirectly pertinent to the list—except possibly because Greece allows all four categories to be seen in the same image—for the reason that it flows into the topographical material of the following chapter and the Ptolemaic map on the verso of the same folio.

In the fashion of an emblematic image, the map that illustrates chapter 17 stands above the title of chapter 18 on the use of Ptolemy's tables (fol. 28r); the map (Figure 12), especially that of the Parisian edition, could certainly pertain to both chapters. The masses of land, like the bodily parts shown earlier, tend to be isolated or, if continental, seen only partially. Cities are marked (Athens, Corinth, Fatras, Modou, Lacedemoine) and so are rivers (Alpheus, Burota). This map figures in implied contrast with the topographical counterpart (Figure 13) set immediately below (and verso) that illustrates the Ptolemaic system. The text advises the reader first to find the place-name of a site or a city in

Figure 12. Pieter Apian, *Cosmographie* (Paris: Vivant Gaultherot, 1551). Map of Greece and text of chapter 18, fol. 28r. Courtesy of the James Ford Bell Library, University of Minnesota.

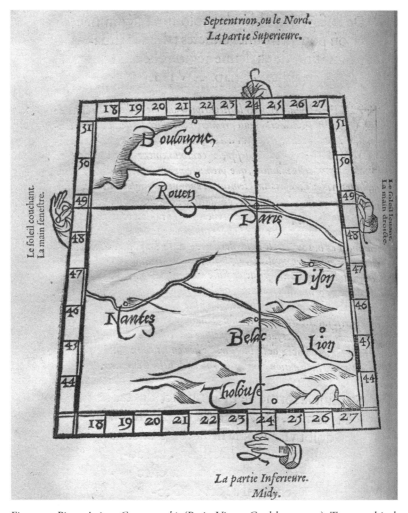

Figure 13. Pieter Apian, *Cosmographie* (Paris: Vivant Gaultherot, 1551). Topographical map of north-central France, fol. 28v. Courtesy of the James Ford Bell Library, University of Minnesota.

the tables of the *Geographia,* locate it in accord with its degree of longitude and latitude, and proceed to establish distances between others by means of two threads, the one perpendicular to the other, to establish the sides on the map. "Le lieu la ou les deux fillets s'entrecouperont [se entretrenchent], sera [le lieu ou] la situation d'icel[l]uy lieu [ou ville] que vous avez desir d'enquerir" (fol. 26r); ". . . filum extendat, locus in

quo fila sese intersecuerint, erit locus aut situs oppidi illius aut habita-
tionis quam investigabas" (fol. 28r) [The place where the two threads
will crisscross or cut over each other will be the place or situation of the
very place or city you desire to locate].

A Spider's Thread

Narrative and cartographical threads are interwoven. The map in the
Latin editions locates Prague as the initial city from which the topography is established, adjacent to Apian's birthplace of Leisnig and to
its neighboring city, Leipzig, where the author grew up and studied.
Other towns, many of them sites in the author's travels, seem to form a
secret map of formative and notable places. A cartographic autobiography is implied, no less, in the listing, below the map, of the longitudes
and latitudes that locate the places that figure in Apian's sentimental
and scientific education: as they are incised in the woodcut, Ingolstadt,
Nuremberch, Monachum, Bavare, Vienne en Austrice, Venize. Quite
tellingly, in the Parisian edition a topographical map of northern
France replaces the projection of Apian's northeastern Germany. Paris,
which supplants Prague, finds its dot on the Île de la Cité, shown vastly
out of scale, that draws attention to the stretch of the River Seine cutting across the landscape. The map puts Nantes at a point inland to
the north and east of the mouth of the river Loire; includes Rouen and
Boulogne to the northwest; Dijon and "Lion" to the southeast; and
"Belac" of the Limousin and "Tholouse" to the south. Yet the list of
place-names and their degrees of longitude and latitude that follow
pertain still to Prague, Vienna, Venice, and the cities in the other map.
It appears that the book underscores its topographical mettle by having Paris, its own site of origin, replace that of Ingolstadt, Antwerp,
Hereford, and other cities in which other editions had appeared. The
map locates the birthplace of the edition itself—but without any textual emendation evincing consultation of the adjacent tables.[17]

Read in view of the final chapter, the most resonant of the book,
the two maps underscore the underlying tension of topography and
cosmography. The hands that animate the map belong to a greater—
and even poetic—network of bodily figures. Like the eye and the ear
seen in the Introduction, what manipulates the strings are at once "any
hands whatsoever" (the left hand is shown holding one thread and the
right another) and extensions of a divine or godlike body, like certain
lands of the earth, that remain unknown. To be sure, they are highly

conventional deictic icons that bring a sense of cardinality and measure to the projection, but they recall immediately the images of hands and legs that elsewhere Apian invokes to make palpable the arts of measurement and of memory.[18] Intermediate or transitional forms that link the abstraction of cosmography to the way in which the world is shown seen and touched, the hands can be implied to belong to readers who would see their own fingers turning the circles of the volvelles that grace every edition of the manual.

The most complex of all is the *Speculum cosmographicum* (Figure 14), "Le miroir du monde" (the cosmographic mirror) that brings the first book to its conclusion and to the threshold of the narrative treatment (drawn from Vespucci) of the three continents and the islands of America. The descriptive definition of the cosmographical mirror extends the inaugural metaphor of the world seen as a portrait of a human sitter. The "semblance" (noted in the vernacular text) of our bodily or corporeal "face or image" to that of the greater world mediates the differences between a totalizing gaze and its localizing counterpart. The reader is invited to behold simultaneously the entirety of the mirror and its reflected parts. A fixed circle of hours, a *limbus in extremo immobilis,* is set below three wheels on the same axis, the first of which is a polar projection of the world, the *speculum orbis* [mirror of the globe] *(quam Mappam vocant),* that the French notes "est communement appellée Mappe du munde" (fol. 28r) [is commonly called a world map]. Above it is *L'aultre,* the other, "pour la semblance des rays ou de la toile que faict l'araigne, est appellé des Arabiens Alhancabut" (ibid.) [for the semblance of the lines or of the web a spider weaves, the Arabs call Alhancabut—*rethe sive aranea nominatur, apud Arabes autem alhancabut* (fol. 29r)—and carries or bears the circles of the zodiac, which is the circle described by the sun or the planets in their course]. Yet the web indeed belongs to the *mappa mundi* and not to the third circle. The central point that on the second map in the earlier chapter had been the umbilical city either of Apian (in most editions) or that of the Parisian edition (for Gaultherot) becomes the North Pole, which, like a human navel, has a very physical presence in the form of a knotted piece of string.[19] The center of the projection is the axial point of contact that requires the publisher to "bind" every circle to the page itself.

Like the circles of the volvelle, the metaphors that describe the moving image are superimposed on one another. The ocular figure of the mirror that reflects a face gives way to that of the gossamer threads of

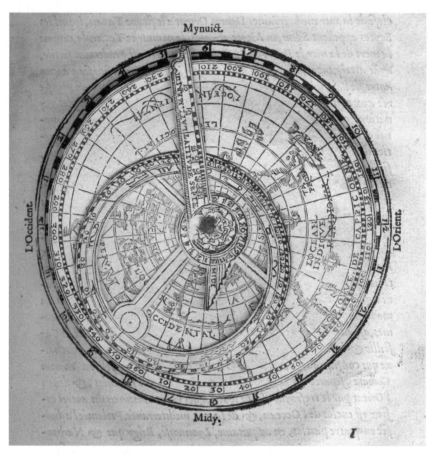

Figure 14. Pieter Apian, *Cosmographie* (Paris: Vivant Gaultherot, 1551). Polar projection of world map as part of a volvelle, fol. 30r. Courtesy of the James Ford Bell Library, University of Minnesota.

the spider web made visible when the reader's hand turns the second wheel and the arm of the alidade, what in the Latin is named *alhidada*. The polar projection turns the world into a metaphor both of an eye and of webbing in which landmasses seem to be held in suspension. In the Parisian edition the continents are finely drawn and outlined by parallel hatching along their coastlines. They are named, it appears, to refer to the physical and cultural geography, announced in the preface that follows, especially in Gemma Frisius's comment about the riches of Peru, indicated on the continent itself. By contrast, the editions from

Antwerp and the north emphasize the expanse of the world's oceans in dark and tightly knit hatchings parallel to the designated circles of the tropics. The metaphors that bind the image and the text address self-location: How does a person go about looking at him- or herself in a vast continental mass of ocean and land? And where will he or she negotiate movement and imaginary travel within the configuration?

In most editions the first of the four propositions comprising the text of the chapter faces the topographical map while referring to its cosmographical counterpart. The reader is advised that to "put any region or city in its place" ("igitur regionem aut civitatem aliquam in Speculo isto locare"), he or she must count the degrees of latitude and longitude in relation to the appended tables. Both maps happen to apply to the first proposition while in the second the circle of hours on the cosmographic mirror will help those who use it to determine exactly where— and, it is implied, who—they are. This is what the rudimentary piece of cultural geography of the second part of the book makes clear by way of the literal thread of its ties to the preceding cosmographical glass. The four continents are studied through the history of their names before being located descriptively. Brief notes about the quality of the land and peoples give way to a detailed treatment of their constituent parts. The farther removed the lands are from Europe, the more their peoples and fauna become savage and monstrous.

A Map of the World and Its Winds

Gemma Frisius's addition about Peru appears to serve as a locational device of its own kind. The praise of the riches of the "fortunate" region that cause it to be worth colonization or exploitation extends to the populations, who are "well instructed in prudence" and in arms and letters, and who are wont to exchange goods. Although they are not yet of Christian belief, it is hoped that they will become so: "speramus, omnique industria, labore, ac diligentia conandum est" (fol. 36v). In many editions Frisius's truncated cordiform map of 1543 (Figure 15) is inserted following the addendum and prior to the listing of the world's regions and provinces, including its cities, towns, mountains, rivers, and islands—everything belonging to the preceding geographical nomenclature. The cordiform projection is *not* correlated to the degrees of latitude and longitude, yet it often does engage cartographical dialogue with material taken up elsewhere. It appears by design to be of loose pertinence: particularly because Peru is one of three points

Figure 15. Pieter Apian, *Cosmographia* (Antwerp: Gregorio Bontio, 1550), Latin edition, truncated cordiform wind map inserted before fol. 31r. Courtesy of the James Ford Bell Library, University of Minnesota.

of reference in America (with "Canibales" to the east and "Gigantium regio" in Patagonia). Apparently of a formal lineage that reaches back to Waldseemüller's map of 1507 and Apian's less accomplished rendering of 1520, it is of truncated cordiform shape but designed to "inspire" or make its readers draw their breath in view of its depiction of the world's winds.

Along the upper right edge at the top sits Charles the Fifth, in the garb and armor of a Holy Roman soldier. He looks back at Eolus, who is astride a great dove, holding a bolt of wind in his right hand while resting his left hand on a massive cloud. The emperor and the god replace the cosmographer and traveler of earlier versions of the same map that featured the portraits and the names of Ptolemy and Vespucci, respectively.[20] The couple find themselves dwarfed by the five wind heads that spew radiating blasts of air resembling megaphones. As the reader moves about the map each of the twelve winds seems to vomit something that would be an attribute: flowers from the temperate winds, pestilent particles from the three death's-head winds to the south, and sleet mixed in the froth from the north. "Charta cosmographica, cum ventorum propria natura et operatione" (A cosmographic chart with the proper nature and operation of the winds): the title and the names of the winds outside of the frame invoke the bodily topography of the world as it is taken up in the fifteenth chapter of the first book, on winds and their virtues.

In editions of the *Cosmographia* after 1544, the fifteenth chapter is added to correlate (or locate) textual description with the thematic orientation of the map. "Le vent," it announces in the Parisian edition, "n'est autre chose qu'une vapeur ou exhalation chaulde & seiche, laquelle s'engendre aux entrailles de la terre, & quand ladicte exhalation sort, si elle est meue & agitée pres les costes de la terre, est appellée vent" (fol. 26r) [Wind is nothing other than a vapor or a hot and dry exhalation engendered in the entrails of the earth, and when the said exhalation is emitted, if it is stirred and tossed near the coasts of the earth, it is called wind]. The four cardinal winds are described and declined in humoral terms, and thus in a figural matrix of the human body. The fourth or north wind "est en soy froid, sec, melancholique, & est comparé à la terre sans pluyes, gardant la santé du corps humain, produysant froidures de nature seiches, nuysant aux fleurs & fruicts de la terre" (ibid.) [is in itself cold, dry, melancholic, and comparable to the earth without rain, guarding the health of the human body, producing dry cold snaps, harmful to flowers and the fruits of the earth]. In contrast, the east wind, born from equatorial areas, "est de nature ardant. Colerique, sec & chaud, attrempé, doux, bening, soubtil, engendrant nuées, & garde les corps en santé, produit fleurs" (ibid.) [is of a burning nature, of choler, dry and hot, tempered, soft, benign, subtle, giving birth to clouds, keeping bodies in health and producing flowers].

The humoral description of the winds bears a corporal presence that the map embodies in its immediate depiction of clouds. They are implied as bowels or intestines attached to the heads of the winds that look upon—and herald with their vomiting breath—the map before them. They literalize "the entrails of the earth" that are paradoxically in the agitated atmosphere around the map. The sole wind head that is in the map, from the Antarctic, is one of the three death's heads shown to bring pestilential vapors to the world. Its position negotiates what the textual description situates in and of the physical world. The design clearly figures in the intermediate or transitional area between cosmographical abstraction and the body of the world in concomitant topographical and biological presence. The editors of the Paris edition and others that date from the 1550s cause the textual description to bear on the imaginative dimension of the map: the fantasies inspired by entrails draw the intellectual gaze *into* the body in the way that the isolated eye and ear of the first chapter invite the reader either to enter the labyrinth near the folds of the lobe or to float on the surface of the pupil fixed on the city view. In the same way, the map seems to engender its space from the area below the crown of the cordiform design, between the god and the emperor, whence the north wind blows so forcefully that the continents seem to spread over the earth's surface from the generative force of the atmosphere. The oriental Indies at the upper left form a narrow strip of an Asiatic continent roughly parallel to the long peninsula of Baccalarum, while, on the other side of the implied meridian, the Scandinavian isles, Europe, and the Tartarian peninsula of Asia push outward.

The force and agitation that typify the map also complement the strong corporal presence that elsewhere is mixed with the practical geometry of location. In the *Cosmographie* the map attests to the virtue of the illustrated comparison of geography to chorography by retaining in its scope the detached bodily parts seen in the first pages. This is not to say that Apian and Frisius's manual includes anatomy in its geography but rather that its tension of abstraction and physicality is determined by the presence of human forms in its various diagrams and projections, as well as in its liminal landscapes. A work of poetry it is not, but as a sum of loosely coordinated texts and images it inspires reflection on how and where the body figures in and is made of the ambient world. It prompts speculation while offering indexical verification of the world and its places. The reader can conjecture that, like the best textbooks for students of cosmography and geography, it promotes as much creative fantasy as geographical truth. In its mix of anachronism and accuracy it accounts for Magellan's voyages; offers a rudimentary anthropology of places and peoples of the four continents; and brings forward a practical trigonometry vital for triangulation. Best of all, it allows the eye to wander from the form and shape of its prose to touch upon the rotundity of its many spherical diagrams. Often, too, at the center of its cosmographic schemata and images of armillary spheres— the site where the "world" would be shown in its balance of oceanic and continental mass—the viewer finds a landscape, indeed the core of a topographical vision. Where the globe is found so also are a landscape and indication of local places. Therein is shown the desire to see the world both from afar and from within, in concurrent detail and totality. The tactile vision of Apian's *Cosmographia* might indeed be what inspires the play of wholes and details in the work, not only of Rabelais, as seen here, but also of designers of emblems and their poets. They are the topics of the three chapters that follow.

3
A Landscape of Emblems:
Corrozet and Holbein

The design of Pieter Apian's manual of cosmography, studied in the preceding chapter, is patently emblematic. The woodcut images that feature ocular figures and their adjacent textual matter invite meditation and action: meditation, insofar as the sight of the world in a polar projection or as a "cosmographic mirror" inspires thought about from where and how the observer can see and touch his or her milieu; action, when correlation of the images with the appended account of the four continents allows readers to extend their reach into the world at large. Now and again, where the textual signs and images redound or crisscross, the woodcuts and their context invite speculation about the measure of the body and of the world in which it finds itself. Some of the combinations can be construed to be riddles or even latent rebuses of the kind found in printed books of hours and manuals of typography.[1]

In this light it can be particularly constructive to study how the topographic orientation of Apian's and other manuals is related to the synchronous development of the vernacular emblem book. The latter has not generally been affiliated with early modern cartography, while it nonetheless tends to call forth questions about perception and sensation of space and place. Without touching on theories of the emblem or tying them to emerging technologies of power when the illustrated book is a fertile ground for creative experimentation and exploitation, the aim of this chapter is, first, to study emblems correlating local and total space and, second, to see what they ask about the nature of extension: When and how is space born and crafted in miniature landscapes in which images of the greater world are found? What is the propensity of the emblem book to ask its viewers and readers who gaze upon it to wonder how and where they are situated in the world, and how does it inspire admiration and disquiet? And no less, in what way does the emblem book inspire a sense of tact and tactility, and of propriety and property, to the degree that ownership and commercial speculation become part and parcel of its landscapes?

The local and general pictures of Pantagruel's world, shown in the second chapter, indicate that, much like Apian, Rabelais was no stranger to incipient emblems or their variants. In both *Pantagruel* and *Gargantua,* local and total images are refracted through the sight and aspect of printed language. Worlds are latent in given letters. As Jean Céard and Jean-Claude Margolin have shown (1984), a genre that he castigated was also the one he espoused. In chapter 10 of *Gargantua* the narrator trashes adepts of the rebus, contra Geoffroy Tory, especially those who imprint their logolike designs on their property or personal effects: mules, pages, breaches, gloves, bedcovers, hairpieces. He finds inept the rebus that identifies the word *espoir* (hope) with *esphère* (the sphere) when it is employed to convey an *imago mundi.* He counts it among these homonymies (including inventive combinations of word and image) that are "so inept, so tasteless, so rustic and barbaric that a foxtail ought to be attached to the collar of each of its users who, further, would do well to have their faces molded in cow manure" now that "good letters" have been restored in France (1994, 29). Those who use the image of the world to signify hope have no sense of the fact that words are not invested in things, cries the author who invests words into things: for this reason, perhaps, the rebus sends the word and image earthward, to the telluric world of fungible matter, while its counterpart, the hieroglyph, aims itself skyward, in the direction of a purer form of virtue. The Rabelaisian emblem or hieroglyph is not an easily decipherable form, as in the equivalence of *espoir* and *esphère,* but of an enigmatic design recalling how its wise practitioners in ancient Egypt wrote not mere letters or images but hieroglyphs.[2] Only they, like the ideal reader of the very words Rabelais puts into print, can understand the "virtue, propriety *[proprieté],* and nature of things figured by them" (1994, 29). Rabelais indicates that he is aware of Horapollo, an author known to Geoffroy Tory in 1529, and that soon appeared in 1543, translated into French by Jan Martin, as a book of emblems.[3]

Yet, in Rabelais, as in books of emblems, abstraction or ideation does not go without matter. His sarcastic quotation of proverbs—"à cul de foyrard tousjours abonde merde" (on a coward's ass shit abounds)—or "homonymies"—"le fond de mes chausses, c'est un vaisseau de petz, et ma braguette, c'est le greffe des arrests" (1994, 29) [the bottom of my breeches is a vessel of peace / farts and my codpiece the mark of a summons / hard-on]—makes visual and aural form coextend. When emblem books seek ideated meaning or an *altior sensus* in common things, or the redemptive tenor of homily, they are nonetheless open to reading that can go both heavenward (in a cosmographical sense of the infinite beauty of the world) and back down to earth (in local practices and places, in the regions where they are used and are seen charted within their form).

The *Hécatomgraphie*

The works of Gilles Corrozet, whether his book of emblems, the *Hécatomgraphie* (1541 and 1543), his collaboration with Hans Holbein that yielded *Les simulachres* (1538), or even his two volumes of *Fables* translated from Aesop (1542 and 1547), are pertinent. Corrozet's talents have been appreciated to be at once entrepreneurial and poetic, and notably invested in the illustrated book. He is an author who is "other" because his typesetters, engravers, and publishers form part of his signature: yet he remains a poet of space, of plotted compositions, and of locale or locational imaging. His verse is composed according to the measure of the page in which it is placed and in rapport with the pictures to which it is juxtaposed. It is often framed, written to resemble epigraphic legends (or subscriptions), and is designed to promote typographic experiment and innovation. He and his editors fit in the context, too, of his properly topographic history and description of Paris, in *La fleur des antiquitez . . . de Paris* (1534), a book that witnessed success and found itself in the company of touristic maps that accompanied it.[4] His edition of Pierre Belon's *L'histoire de la nature des oyseaux* (1555), an ornithological survey, a natural history, and a taxonomy, is built along the lines of an emblem book.[5]

In this broader context, how the images of the world "work" in the emblems of the *Hécatomgraphie* bears on the sense of cosmography and topography witnessed in Apian and Rabelais. Wherever an emblem designates the globe or world, it deploys the seemingly "inept" association of the *sphere* with *espoir,* and in doing so it engages issues of perspective and of place wherever a sign of a greater world is seen within the confines of a finite and minuscule frame. His preface, whose bold superscription in cul-de-lampe format and point size progressively reduces from the first to the third line, announces its origin ("Gilles Corrozet / Parisien, aux bons / espritz et amateurs des lettres"). The decasyllabic poem of seventy-two lines declares that what he brings to light must issue from shadows or has been disguised so that he can do with it what he will. In the same gesture he clarifies and he occults; he makes the eye aware of the way it sees and reads, and of

> Ce qui fut dict des gens de bon sçavoir,
> Le desguisant, pour mieulx le faire veoir
> A l'oeil de tous, comme on faict par raison
> De vieulx mesrien une neuve maison. (fol. Aiiii r, lines 26–29)[6]

> [What was said of people of keen insight
> Disguising, better to bring it into sight
> To the eye of all, as is done with reason
> When an old shed is turned into a new mansion.]

The eye given to see and to see and look otherwise, to produce a new and finished property, perhaps like that of the smooth and cool interiors seen in *Les blasons domestiques* (1539), his emblem-blazons about households and their furnishings. *L'hécatomgraphie* aims to make the ocular quality of the images and printed writing a pleasure to behold.

> A fin que soit plus clairement monstrée
> L'invention, & la rendre autenticque,
> Qu'on peult nommer lettre hieroglyphique. (fol. aiiii v, lines 58–60)

> [So that more clearly be shown
> The wit and to make it authentic
> The letter may be called a hieroglyph.]

The enigmatic writing will configure animals in its implied design, much like a book of fables, to convey parables concerning vice and virtue. Thanks to the book's illustrations, the preface asserts, a public of illustrators, woodcutters, painters, jewelers, weavers, and ceramicists can "find fantasy" vital for their creations (lines 65–66).

Hope in the Sphere

The tightly organized frame for the illustrations enhances the domestic, practical, and discrete ends of the book. The volume contains one hundred "emblems, authorities, sentences, apothegms of learned authors, including Plutarch and others." A totalizing form, a world unto itself, the book is composed of one hundred allegorical sites, scenes, and landscapes. The sign of the globe appears only five times, in close proximity in emblems (not numbered in the text) 41, 46, 48, 49, and 58. In emblem 41 (Figure 16), whose motto is "Esperance en adversité" (Hope in adversity), printed in the upper area of the window, the *inscriptio* or woodcut picture portrays an allegorical figure, a female dressed in billowing folds, bearing a globe on her right shoulder while wading across a body of water and approaching a shore to the right. The scene seems literally knotted at its center. From the side of the woman's navel extend the pleats of her toga and a snakelike ribbon that describes a circle at her waist. Its other end wafts in the air like a caudal fin and is matched by another behind the left side of her back. The two tufts of fabric bear slight resemblance to wings found on Mercury's sandals or those that would be attached to the upper back of an angel. "Esperance" finds herself in the midst of an almost anthropomorphic landscape in which rocky outcroppings, replete with sprigs of brush, are on either side. In the background a low mountain range encloses the body of water whose ripples are depicted by tight parallel hatching. A foliate tree on the left has as its counterpart a palmate species on the other. The woman finds what seems to be a welcome closure in the space as she bares her left leg to reach the shore on the right. On her head—what Rabelais in his diatribe against the rebus calls "the bonnet's mold, the wine pot, as my grandmother used to say" (1994, 30)—she braces an armillary sphere on her neck and right shoulder, her right hand resting on the north pole to hold it erect and make visible its lines of the tropics, its colures, and the ecliptic band.

In the configuration of the emblem the woman "becomes" allegorical through the textual gloss. Not described as one of the Pauline trio of "Faith, Hope, and Charity," she arrives at her name, Esperance, through toil that the poem indicates by calling the body of water "the sea of adversity": "En ceste mer avoir nous fault / Bonne esperance sans default" (lines 15–16). [In this sea we have a need / without doubt good hope to heed.] It continues in describing the globe:

> Ceste esperance est figurée
> Sus la Sphere bien preparée:
> Ou est paint chaque element,
> Et le tournoyant firmament,
> Et les cieulx. Pour nous faire entendre
> Que là hault nostre espoir doibt tendre. (lines 17–22)

Figure 16. Gilles Corrozet, *Hécatomgraphie* (1541), emblem 41, "Esperance en adversité,"
on fol. v facing fol. G r. [Typ 515.43.299] Houghton Library, Harvard University.

[This hope is figured here
On the well crafted world and sphere:
Where is painted every element,
And all about the twisting firmament
Are the heavens. Thus to comprehend
To the skies our hope must tend.]

By virtue of carrying the armillary sphere, the woman, like Atlas in the future folios of his name, acquires the attributes of the allegorical figure by way of the association of hope *(espoir)* with the armillary sphere. Where the *subscriptio* enjoins the reader to look skyward, the image offers a horizontal or scenographic view of the landscape. The ambient cosmos, described here in Ptolemaic terms (the firmament turns about the earth at the center of the sphere), is countered by the picture that directs the gaze, not upward, but across the local landscape, over the sea—"full of bitterness"—where the winds trouble it, "wave against wave redoubled" (lines 4–6). Yet in the image the waters seem calm and the vegetation slightly reminiscent of the Holy Land. The perspective draws the eyes from the periphery to the knotted folds at the woman's midsection, but the undulating lines then move outward, to the left and right, to join those that depict the rocky contour of the landscape in the background. A world beyond the frame of the emblem is intuited within the struts of the armillary sphere, while the view of the landscape and its design aim the construction toward a divided origin, not just of text and image *(emblemata, sive epigrammata,* as Alciati had defined the emblem), but of two different but inseparable ways of contemplating and apprehending space: one, as total and infinite; the other, as anchored in specific places, which here is the body of the woman who is earning her status as an allegory doubling or becoming the icon she bears. The "world" is perched on Hope's shoulders, too, to suggest that the eye move up and out of the frame (perhaps toward the thistlelike rose in the shield bearing some resemblance to a heart—the device of the *coeur rosé* or rosied heart of "Corrozet"—under the voussoirs of the arc above). The ocular passage noted in the last four lines is also comparable to those found in emblems of death: we must look skyward in hope of salvation and remind ourselves "D'avoir espoir d'aller un jour, / Faire là hault nostre sejour" (to have hope to go one day / Up there to make our stay). In its relation with folds of cloth and the landscape, the world becomes a cursor with which the eye can sense itself moving about the landscape and across its frame.

The world that appears in the landscape of emblem 46, "Le monde instable," carries greater cartographic latency. It displays a gentleman, his left hand holding a staff and his right arm extending to display the orb of the world, standing on a floating island amidst scalloped hatching signifying the torrents of the seas. Behind is seen a horizon of a *terra firma* and to the left a floating city (which indeed approximates the view seen in Pieter Apian's similitude of geography and topography in Figures 2 and 3). The quatrain below suggests that the man's staff is an imperfect rudder that cannot guide the floating island to stability or safety:

> Le monde en une isle porté
> Sur la mer tant esmeue & rogue,
> Sans seur gouvernal nage & vogue,
> Monstrant son instabilité.]

> [The world borne in an island
> On the haughty, unsettled sea,
> Without a sure rudder swims and wafts,
> Showing its instability.]

The elements of an *isolario* are made clear in the subscriptive poem in which "mundane virtue" is seen floating as might the island seen in the image, "unaware of the next wave / that only seeks to sink it." The gist of the device inverts the sense of the forty-first emblem while rehearsing that of the emblem that follows two folios below.

The forty-eighth emblem, "Trop esperer decoipt" (Too much hope deceives), appears crafted to mediate and to establish a serial pattern with emblems 41 and 46 (Figure 17). Here an old man—and not an intrepidly resourceful woman—stands in similar waters, gesticulating with his right arm, pointing his index finger skyward, while with his left arm he vaguely points at the shore where the armillary sphere is set in a fire to illustrate that it is "without hope." The reader wonders if he is shown telling the lesson written on the opposite page, ventriloquizing, as it were, Corrozet's verse in which excessive hope is shown to be cause for correction. An initial reading is that he stands in the troubled water "in great danger" and thus is led astray from reason. His beard and pensive air suggest that he is a sage, but in standing in the rushing currents (emphasized in the image), he may be a fool. The unbound and mobile geography of speech and place in the relation of the image to the text promotes a double reading.[7] The contrast of the foliate and palmate trees has as an analogy in the latent "moralized landscape" in which antlerlike

Figure 17. Gilles Corrozet, *Hécatomgraphie* (1541), emblem 48, "Trop esperer decoipt," on fol. v behind fol. H r. [Typ 515.43.299] Houghton Library, Harvard University.

branches are over a rocky cliff on one side, and a full and shaded grove of pines on the right stands adjacent to a stenographic sign of a church in the background, not far from the man's face. The wafting folds of the man's cape resemble the hills and valleys of the ear that Apian had placed in his similitude of geography and topography (Figures 2 and 3).[8]

A sense of movement and montage is afforded when the latter image is superimposed over the former: a metamorphosis of space takes place by way of the displacement of the globe from one emblem to another. Emblem 49 literally carries the icon of *Espoir* from one site to the next (Figure 18). "Esperance conforte l'homme" (Hope comforts man) brings Fortune, the crucial figure and theme that will close the volume in the hundredth emblem, into the picture at the middle. A traveler who bears the heavy Wheel of Fortune on his back walks ahead, his face (somewhat in the style of a thirteenth-century Gothic figure) looking at the path his feet have followed. He gains stability with a walking stick that he firmly holds in his right hand. An armillary sphere is placed in the middle of the stick. Held so as to touch the ground, the cane has the locative virtue of helping the viewer find himself in the world at large, at the very least because, like the snail of the twentieth emblem (Figure 1), looking backward, he is blind to what is before him. The armillary sphere is shown offsetting the correlative shape of the wheel on his back that is its geometrical complement and counterpart. The pilgrim who ventures into space is armed, nonetheless, with a locative device, the cosmographer's version of a global positioning system. Now, in direct contrast to emblem 41, a palmate tree is on the left and a foliated arbor on the right while, in the background, the view of a fortress on a hill suggests a distance that the wanderer has covered while, as he says (and here the voice of the quatrain is attributed to him):

> Si fortune soutiens & porte,
> Qui m'a faict ung tour inhumain,
> Je tiens esperance en la main,
> Qui me conduict & me conforte.

> [If I hold up and carry fortune
> That plays an inhuman trick on me:
> I hold hope in my hand,
> That leads me and comforts me.]

Further, the cane-and-sphere is juxtaposed to the tip of what appears to be a sheath or a sword that is strapped to his left side. The pommel is figured as a circle with a dot at its center, located before the wheel of

Figure 18. Gilles Corrozet, *Hécatomgraphie* (1541), emblem 49, "Esperance conforte l'homme," on fol. v before fol. H r. [Typ 515.43.299] Houghton Library, Harvard University.

fortune, next to the view of a fortified town in the background. Before it is seen as such, the tip of the sword can be construed—that is, correctly mistaken—to be a prosthetic crutch with which the man can limp along. The reading viewer wonders whether the figure is bearing not only the moral of the emblem but also an implied narrative thread that is woven by the recurrence of objects and forms in the successive images and the variety of possible readings. The pilgrim is the mirror of the errant wanderer, indeed the reader, whose eyes move at once through the details of the image, the gloss of their printed letters and words, and the enigmatic window frames in which they are presented.

When the armillary sphere appears for a fourth time in the *Hécatomgraphie,* in emblem 58 (Figure 19), the reader wonders if it belongs to a narrative filigree in which thematic and visual strands are interlaced or if, by hook or by crook, they just happen to be where they are or are the figments of the emblematic design.[9] As in Freudian dream work, the design can be at once latent and manifest. What was not yet Hope *(Esperance),* as known in Saint Paul's first letter to the Corinthians (a text crucial for the promotion of Gallican iconography in the 1530s and 1540s), in emblem 41 is named as such when the same globe reappears.[10] In a different but less ornate surround, the motto "Foy, Charité, & Esperance" stands over a frame of identical aspect ratio (1.25:2), in which the three virtues are disposed much as the trees, landscape, and central figure of the forty-first and forty-eighth emblems. The quatrain that underwrites the image makes the point saliently clear. An elegantly dressed man in the center holds an armillary sphere in his right hand, while his left clasps a rope, tied to resemble a fleur-de-lis, to which it is attached. Faith is depicted by two disembodied hands that shake each other. "Charité par feu est escripte" (Charity is written by fire), while "Esperance par Sphere est dicte" (Hope is stated by the Sphere). To the left of the coiffed gentleman, a four-legged creature (would it be an iguana?) is perched atop a high rock from which, below, emerges either a snake or possibly an egret or a heron, its neck curved like the cord that holds the shaking hands. To the gentleman's right another rocky crest finds at its top an animal that seems to be crawling up a tree that, like the branches on the other side, is lopped off by the upper horizontal edge of the frame. The landscape is defined by shrubbery in the foreground, a hillock in the middle, and signs of habitation in the back.

The disposition begs for inquiry. Hope is in the center of the image though not in the text of the motto. Fire, the attribute of Charity, is nowhere or at best difficult to discern. The explicative poem on the

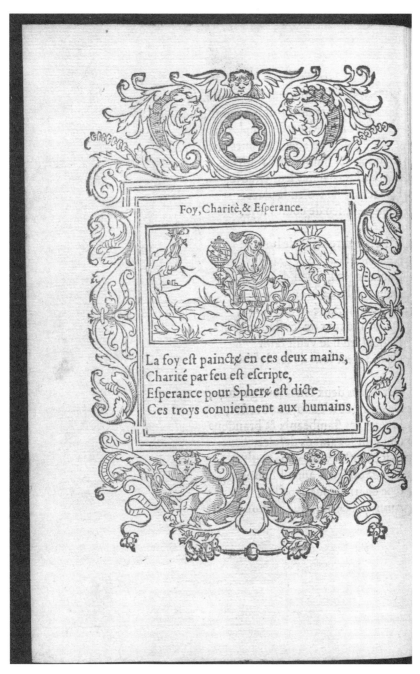

Figure 19. Gilles Corrozet, *Hécatomgraphie* (1541), emblem 58, "Foy, Charité, & Esperance," on fol. v facing fol. Iii r. [Typ 515.43.299] Houghton Library, Harvard University.

(recto) folio opposite the image is so homiletic (or sanctimonious) that it offers no convincing solution to the hieroglyph. The first stanza of the decasyllabic poem of forty-two lines seems designed to have Christ, in lowercase roman majuscule, flow over the end of the fourth line and be suspended in a space that draws attention to the word and its isolation. Faith is, of course, the topic of the first stanza, Charity the second, and Hope the third. In the last it is said that to hold to one's promise, words of assurance are needed. Only then can we take hope in arriving at "this great altitude" (line 20). If we believe in "Iesus," we will have a spur *(esperon)* of hope to inspire us and give us force.

> Et ne craignons que l'ennemy nous blesse
> Si en Iesus, nostre Dieu esperons:
> Car esperance est un des esperons,
> Qui nous induit, & donne hardiesse. (lines 21–24)

> [Let us not fear that the enemy wounds us
> If in Jesus, our God, we hope;
> For hope is one of our spurs
> That induces and brings us courage.]

Hope is iconized both as a globe and, by way of visible homonymy, as a spur—which has no correlate in the image. The enemies would now be the creatures that emerge from or dissolve into the earthly landscape for those persons in whose dreams and diurnal vision monsters awaken. The woodcut plays on the foibles of sight in view of a world in the hands of a forthright and, it is presumed, an engaging entrepreneur. The sphere that had become a virtue in the earlier emblem is now implemented for practical ends, in the here and now, where the possessive individual might be found. Economy and topography are yoked together.

Emblems Compassed

They are also found in another emblem in which geometry, the art of cosmic and practical reason, is shown arrestingly in emblem 94 (Figure 20), "Entreprendre par dessus sa force" [Going beyond one's means], in which two landscapes are superimposed (somewhat like the two city views Magritte makes famous in his *Promenades of Euclid*), and in a grouping of forms similar to what obtains in emblems 41, 44, and 58. A broken circle contains a gigantic pair of dividers, one of whose two arms is shattered. It stands erect, however, in front of a hilly landscape whose depth of field is marked by the broken arm lying on the earth,

Entreprendre par deſſus ſa force.

*Celluy qui ſon eſprit efforce
Et veult plus qu'il ne peult côprendre,
C'eſt comme qui veult entreprendre
Oultre ſon pouoir & ſa force.

Figure 20. Gilles Corrozet, *Hécatomgraphie* (1541), emblem 94, "Entreprendre par dessus sa force," on fol. v after fol. Giiii r. [Typ 515.43.299] Houghton Library, Harvard University.

set adjacent to a rocky ridge in the background, a shrub in the middle, and a building (sporting a tower) in the back. Outside the circle, to the right, a densely foliated tree, perhaps an oak, is opposed to an out-building in the middle ground that seems to be part of a fortress or a battlement, a lower level opened by a porte cochere. The enigma, like many of the emblems that display practical tools without the presence of humans to manipulate them, turns the broken dividers into an iconic memory image. The quatrain coaxes the meaning to emerge from obscurity: those who force their minds and seek more than they can comprehend resemble whoever undertakes a project beyond his or her power . . . or who lacks the "compass" to do so. An alert eye is needed to discern the small break in the arc between the positions of noon and two o'clock before the poem in subscript calls attention to the gap. The presumptuous or overenterprising individual is said to "resemble a compass extended to describe a circle, that is pressed so much that it breaks: and the circle *[rondeau]* in fact already begun is left imperfect," proving "[in] what way the master does not do what he claims" (lines 14–18).

A broken compass and a broken circle: the former within the latter (the cause within its effect) displays the circular spring at the juncture of its two legs, thus doubling the outer circle. The element essential to the construction of the topographic view, the dividers, stands within the scene that in itself is doubled. However it may be broken, the mode of production of the emblem is shown within the emblem, lending to it a sense of autonomy in which the absent hand that has drawn it is integral to the image and is sturdy enough to bring the analogy to completion. Like that of many of the other emblems, the landscape bears strong allegorical traces. The thick and dense leaves of the hardwood to the right stand in bold contrast to the stunted vegetation to the left. One is complete and the other merely begun, *laissé imparfait* (line 17) [left imperfect]. The countryside conveys the allegory insofar as the conventional division of the "rich" and "poor" or fertile and sterile areas plays into the visual rhetoric of the design. The landscape as abstraction, seen as a form almost encircled by the instruments responsible for its production, invites further speculation.

Why the play of perfection and imperfection where geometry is concerned? The open space of the circle accounts paradoxically for the way the emblem is made, as a composite unit of drawn and engraved parts that cannot entirely be joined to one another without gaps and fissures in the design. This emblem is clearly fashioned from four separate blocks that comprise the ornate surround. The entablature at the top supports two grotesque mascarons growing from spiraled stems that end in the face of fishes or dolphins attached to other floral motifs. It is supported by two column blocks decorated as faces (in profile) of two men of stern appearance looking outward, one glumly (to the left), the other pensively (to the right). Their facial features grow from coiled acanthus leaves that end (or begin, depending on how the design is read) as either an array of leaves or a console. A gap is opened at their upper and lower edges. The two blocks seem to float below the entablature above and the lower cornice to whose lower edge are attached grotesques clasping fruit sways and festoons. Although the ensemble has the appearance of an ornate window decoration (in the style of Hugues Sambin and any number of other architects), its broken junctures, due to the way the blocks are set for the printed page, lend to the image on first glance a perfect form, such as the circle, but only before a critical gaze notes how it has been assembled. The same holds for the space in which the image, the motto, and the subscriptive quatrain are placed. They are set in the "machinery" of the emblem, but they are autonomous and can be replaced by any images or lines of type that can fit within a nearly square form (2¼″ x 2⅛″). The motto thus arches or circles back, like the design on the periphery, to reflect some of the constraints made clear in the production of the emblem. The moral lesson aims toward an area of tempered conduct and good judgment but also in a material direction, to the very space and place of its own creation.

They are especially marked in the first and last emblems implied to be the cornerstones of the volume. The one hundredth is significant for the way it serves as both an emblem and a colophon to the *Hécatomgraphie* (Figure 21). "A qui Fortune en donnera" (To whom Fortune will bring fortune), in one of the simpler of the four types of frames used throughout the collection, is set above a crowded image of the general perspective seen in emblems 41, 58, and elsewhere: in the center sits a man whose outstretched arms appear to be dropping objects into two glass containers. To his right, and in a staggered arrangement, three men, the farthest seated in an ample armchair, look at what the central figure is doing. A similar arrangement is seen to his left, except that the man is seated adjacent to another who holds a clarion. The face and robust body of the central figure are made manifest by the window (depicted in dark parallel hatching) at the end of a thick wall shown in the depth of field. The quatrain below does little to explain the staging, except that the sense of a whole, a totality of the world, is intimated through the keynote repetition of *Tout* at the corners, as it were, of the two cornices of the poem:

> Tout ce qu'ont dit les anciens
> De Fortune, & sa liberté,
> Qui donne des maulx & des biens,
> Tout est icy representé.

> [All that the ancients have said
> Of fortune and her liberty,
> That confers fate good and ill,
> All here is represented.]

Where, however, is Fortune? Neither the goddess nor any of her attributes are visible, as they are in emblem 40, where she stands naked on a dolphin and a globelike buoy in the sea; or, too, as she does in emblems 49 and 87. The adjoining poem of thirty-four lines—the longest of the entire collection—spills onto the verso of the last folio. Another space, less allegorical than real or local, indicates why: Fortune is best figured by a game, "Qui de present est joué dans Paris, / Nommé la Blanque" (lines 3–4) [Which today is played in Paris, / named *La Blanque*]. Some are said to win and others to lose. In one box are the "names and devices" of those who have invested copiously in the wager, and in the other are white signs that are their facsimiles, "among which are set the awards for the gracious and propitiously lucky ones [*remembrants* (1543) or *rencontrans* (1541)] who reap jewels, stones, bowls, chains, gold finery, necklaces, rings, cups, belts, and other goods [*biens]*" (lines 18–23). The weight and value are marked on some of the written signs. The man between the boxes is blind. When he happens upon [*rencontre]* signs that match the reward of good fortune, the latter goes to the one whose device is encountered [*rencontré]*. The game is thus named because the losers are left blank.

This variant on bingo portrays Fortune *in Paris,* the site of the publication of the *Hécatomgraphie,* and the prizes take the form of the object for which, as Corrozet stated at the outset, the emblems are destined. Fortune is materialized in the production of the image in the place it chooses to call its origin and to authenticate through a local practice. The first poem underlines how much—paradoxically—"all" that the classics have said of fortune is "all" shown here. The landscape of the

Figure 21. Gilles Corrozet, *Hécatomgraphie* (1541), emblem 100, "A qui Fortune en donnera," fol. Oiii v. [Typ 515.43.299] Houghton Library, Harvard University.

ancients gives way to that of contemporary time, no doubt (and coyly
so) to move the moral of "all" of the book to the colophon below the
author's *devise,* his sign, his apothegm, "Plus que moins" [More than
less], set above the highly locational words that can be taken as a terminus or a projective form (where *fin* is both an end and an ideal goal), in
the words noting that the book was newly printed by Denys Janot "in
Paris, in the New Notre-Dame Street, at the sign of John the Baptist,
in front of Saint Geneviève-des-Ardents." Like the symbolic snail of the
twentieth emblem (Figure 1), the book and its process emerge from and
recede into a local dwelling.

The art of the book is found in the connections and breakages that
form an itinerary, a *via rupta,* that can be led from one emblem to another. The strong degree of parataxis within each emblem and between
their sum allows for readings that take account of the ways the viewer
can see how the work is being read, if not even like itself as a com-
modity or *bien,* according to the spirit of different occasions. How one
looks at or speculates on its enigmas, that is, how readers can account
for or plot how they locate themselves in respect to its form and origin
becomes a commanding and highly dialogic condition—the event—of
its production of meaning. Which the inaugural emblem makes clear
when the author appeals to the reader more directly than anywhere in
what follows: "Vous qui entre les gents parlez" (line 1) [You who speak
among people], "Vous qui plaidez es justes causes" (line 9) [You who
plead for just causes]. Through these words the book aims at the reader
whom it concurrently locates, names, and targets by way of ricochets of
analogous images, ornate decorations, and secretive or enigmatic verse.
Not by chance does the image below the motto of emblem 1 (Figure 22),
"Parlez peu & venir au poinct" [Speak little and get to the point], display an archer pointing a crossbow at a target nailed to a flat and broad
surface.

His aim bears analogy with what is shown in the hundredth emblem.
The simple quatrain above the image seems to explain it as directly as
the archer fires at his target—"just as" he who shoots and lands his
arrow directly in the bull's-eye will gain esteem and praise—in order
to lead into the moral of the longer poem, "so then" do speakers who
use words directly and cogently for their own ends. But the emblem indeed "speaks little" or in a mute way. The image depicts three men, the
most prominent being he who aims at the target; behind him a second
contestant arches his crossbow with a windlass, and behind him a third

Parler peu & venir au poinct.

*Celluy qui le mieulx tirera
Droict au but & plus prest du blanc,
Son coup sera estimé franc
Et la louenge en recepura.

Figure 22. Gilles Corrozet, *Hécatomgraphie* (1541), emblem 1, "Parler peu & venir au poinct," verso of folio facing fol. B r. [Typ 515.43.299] Houghton Library, Harvard University.

figure looks at the target at the right of the image. Three different moments of time are shown in the same space: the verge of the act of firing, the preparation for that moment, and a visual assessment after the fact. The men's sight line that moves *across* the image is countered, implicitly, thanks to the ocular quality of the bull's "eye," by the reader's eye that draws its aim onto the scene and the locale. The specific nature of the topic, "target practice," shown as an exercise and as a game in a given place, becomes, it can be said redundantly, emblematic of the emblem.

Simulachres & histories faces de la mort

The sense of place and space that Corrozet puts forward in Denys Janot's editions of the *Hécatomgraphie* may have its most pertinent and telling expression in an earlier and formative masterpiece, his *Simulachres & histories faces de la mort,* published in Lyons "Soubz l'escu de Coloigne" (under the escutcheon of Cologne) in 1538. One of the greatest of proto-emblem books in design and execution, replete with forty-one woodcut images by Hans Holbein, the book is of a crisp and open, indeed lively and spacious aspect, of irony and of wry wit where the theme of death would otherwise impose sententiously dull lessons. In *Les simulachres* it can be said not that to philosophize is to learn to die but, rather, that to treat death with wit is to learn how to live. And like the labor of the *Hécatomgraphie,* its paratactic arrangement of recurring details within the images and their subscriptions encourages a reading that cues on the space it creates as it goes ahead and about, backward and sidelong, and in twists and turns.

Of a deceptively simple arrangement, it figures its sum of scenes that relate death to appropriate quatrains drawn from the Bible.[11] The Latin texts and their location in the New and Old Testaments are set in superscription.[12] Below the images (1⅛″ x 2½″) are Corrozet's translations in octosyllabic quatrains. The book is composed of woodcuts that follow, first, a prefatory letter to the abbess of the convent of Saint-Pierre-de-Lyon, Jehanne Touzelle, from Jean de Vauzèle, who describes himself as *un vray zèle* (a true zealot) and, second, a piece titled "Diverse Depictions of Death," said not to be drawn or imagined, but "extracted from Scripture, colored by Ecclesiastical Doctors and adumbrated by Philosophers" (9/fol. B r). The success of the book, which saw eleven editions from 1538 to 1562, can be attributed, notes editor Werner Gundersheimer, to Hans Holbein's art of handling the theme of death. He draws on the tradition of the *danse macabre* and the *memento mori.*

"One might say that Holbein's work synthesized these two strands of representation, placing the familiar social types of the Dance of the Dead into individual vignettes of the 'memento mori' type" (1971, x). Holbein transmits the "overriding idea of the omnipresence and universality of death" (xi), taken to heart in the sixteenth century, when humans believed that their immortal soul would submit to the final judgment of God. The collection belongs more to the theme of the *ars moriendi* than that of the *carpe diem*. To mobilize these themes Holbein sets symbolic devices or "props" into the field of the image, notably an hourglass, drapery, musical instruments, and the attributes of each of the orders shown (the bishop's crosier, the emperor's diadem, the laborer's plow, the mendicant peddler's back basket). Gundersheimer observes further that the skeletons have far more wit than their victims, and that they are placed in a "variety of landscapes, backgrounds, human types, emotional expressions, physical positions, and styles of dress and architecture," thus turning the "thematic limitations of his subject" into "a vehicle for his immense pictorial range" (xiii). In his jeweled introduction the editor moves from topical analysis, inherited from a copious bibliography of the representation and presence of death in everyday life in the early modern age, to a nascent topographical counterpart by which pictorial style inflects the variety of things noted and shown in the woodcuts. In a strong sense Holbein's signature becomes locative or locational.

His renderings of the armillary sphere, the cosmographer's object of speculation par excellence, figures among the links that concatenate the work and infuse it with an arresting sense of topography. The sphere appears only once as an attribute of the astrologer (Figure 23, woodcut 25, 42/fol. F v), one of the first in the series of images of people of known professions, following a majority of themes devoted to personages representing the spiritual and temporal orders of a kingdom. Seated in his study by an open window, in a robust armchair whose spiral newel is beyond the grasp of his right hand, the astrologer looks stymied in thought when gazing upward to view the large armillary sphere that hangs over his desk much as would a chandelier. Thanks to the folds of a white curtain that hangs from the cornice of the upper edge of the woodcut, the components of the sphere clearly emerge from the background. The terraqueous globe in the center is encircled by the colures, the equator and tropics (while the position of the broad ecliptic, on which are drawn lines approximating the signs of the zodiac,

Indica mihi fi nofti omnia.Sciebas quòd
nafciturus effes , & numerum dierum
tuorum noueras?

I O B   XXVIII

Tu dis par Amphibologie
Ce qu'aux aultres doibt aduenir.
Dy moy donc par Aftrologie
Quand tu deburas a moy uenir?

Figure 23. Hans Holbein, *Les simulachres & histories faces de la mort* (1538). "The Astrologer,"
fol. F v. [Typ 515.38.456] Houghton Library, Harvard University.

resembles the latter).[13] On the top of a squat round table—almost seen from an anamorphic perspective—in front of which he sits are a bookstand and an open folio volume, placed behind a sextant and next to a bowl, possibly a cut or *coupe* from a globe on a base of three struts. The astrologer looks upward almost blindly while his right foot touches the lower edge of the enclosing frame. An intrepid male skeleton holds a death's-head in front of him. He remains oblivious to its presence but with his foot signals that it is placed within the frame of an image. The deathly grotesques that decorate the chair seem aghast at what the skeleton presents. The homily of the image would dictate that the diviner of the heavens is blind to the pressing fact of death, and that he would do better to look at the skull that belongs to the char and soil of this world, instead of attending to the ether of the planets or the schemata that approximate their orbits.

Details reveal a complex tension of geography and topography. The surveying instrument is shown as a spherical triangle set under the spine of the book. The support on which it rests displays an arm that composes a right triangle, the hypotenuse identified as the edge of the stand and the other side of the surface of the table. The support draws a vertical line that goes, with the edge of the curtain in the background, all the way to the top of the frame. The arc of the spherical triangle is parallel to that of the table. And somehow the spiral of the arm of the chair engages a twisting motion that can be *seen* in the way, first, the heavens would turn about the planet Earth at the center of the armillary sphere and, second, in the rotational design of the table and the instrument insofar as they are yoked to a pun (*amphibologie* in Corrozet's gloss) by which, when the astrologer meets the death's-head, "the tables are turned."[14] Bright light is figured flowing through the open window, "illuminating" the blind astrologer along a sight line identical to the direction of the gaze of the death's-head and the skeleton that bears it. The play of light and shadow in the globelike forms on and below the table enhances the effect of rotation and the concurrent presence of vision and blindness along with the alternation of day and night.[15]

The measuring instrument on the table does more than enhance the effect of rotation. Clearly the astrologer is not just a cosmographer, a student of the heavenly bodies meditating on the armillary sphere. The spherical triangle suggests that he is also a surveyor, a professional whose gaze can extend to limited areas; and the compendious book on his table could be of the order of Sebastian Münster's *Cosmographia,* begun in 1530, concurrently with the *Simulachres*. The man who studies both the heavens and the earth belongs to the evangelical world of Erasmus and Rabelais, to those who follow the instructions of Psalm 104: "to recognize the infinite diversity of the creatures who populate the surface of the globe inasmuch as this diversity reflects the glory of God" (Besse 2003, 162). The figure of death confronting the astrologer can be taken to mean, by way of another rustic proverb, that the scientist ought to look at the things of this world and the shadows cast on its land and water. Hence the interest of the spiral at the end of the arm of the chair and the decorative legs of the table: the spiral spins off a wheel of five radiating spokes; the crying mascarons below seem made of living wood; the tibia of the skeleton's left leg is both a tibia and a strutlike form mediating the astrologer's leg and the "legs" (or birds' necks) of the table. In all events the cosmography affiliated with the astrologer is enhanced by the details in the image that call attention to what it means to look closely at the world—topographically—when time and space are contingent.[16]

On Top of the World

The only other *imago mundi* in the *Simulachres* is in the penultimate and decisive cut of the Last Judgment, the fortieth that caps the procession. It follows the emblem of the young child (Figure 24), beckoning to its mother with its right arm, which touches some billowing folds of smoke turning about in a coil (so as to make clear the presence of a gibbet in the pattern of the beam of the barn). The mother toils over a fire in the dilapidated setting while a skeleton holds the infant's left hand and readies to carry it off, out of frame, to the right: toward the image of the Last Judgment in the next folio recto (Figure 25). The infant is being removed from a local and highly specific milieu to a site where a vision of the world and its spheres commands the image. A great armillary sphere, an implied rebus of Hope, occupies the center of the frame. A tired but robust Christ, his upper torso exposed, sits in the guise of an emperor at the top of the world. Radiating hatching depicting shafts of light and bowel-like clouds form a halo behind the deity's head. His left hand in trompe l'oeil perspective reaches out, as if into the image, to display the scar left by the nail of the Crucifixion. The great gallery of various orders of resurrected souls cheers Christ's presence over the world below. The great arc that touches the upper edge of the world on which he squats (he does not seem comfortable in the position or at the thought of the duties he must perform), and on which his robes are

Homo natus de muliere, breui viuens tempore
repletur multis miserijs, qui quasi flos egre=
ditur, & conteritur, & fugit velut vmbra,

IOB   XIIII

Tout homme de la femme yſſant
Remply de miſere, & d'encombre,
Ainſi que fleur toſt finiſſant.
Sort & puis fuyt comme faict l'umbre.

Figure 24. Hans Holbein, *Les simulachres & histories faces de la mort* (1538). "The Child,"
fol. Giii v. [Typ 515.38.456] Houghton Library, Harvard University.

Omnes ſtabimus ante tribunal domini.
ROMA. XIIII
Vigilate,& orate,quia neſcitis qua hora
venturus ſit dominus.
MAT. XXIIII

Deuant le tróſne du grand iuge
Chaſcun de ſoy compte rendra,
Pourtant ueillez,qu'il ne uous iuge.
Car ne ſcauez quand il uiendra.

Figure 25. Hans Holbein, *Les simulachres & histories faces de la mort* (1538). "The Last Judgment," fol. Giii v [fol. Giv r]. [Typ 515.38.456] Houghton Library, Harvard University.

draped, forms, with the sphere, the partial shape of an immense eye that gazes directly upon the viewer.

The armillary sphere, drawn in great detail, holds the earth in suspension within the heavens in the volume of the greater sphere, which is girded by its lines of latitude and longitude and the ecliptic band on which several signs of the zodiac are visible (reading from left to right and bottom to top: Aquarius, Taurus, Cancer, Libra, and Sagittarius [?], clearly in a mixed and motley sequence). About and around the sphere, looking and gesticulating skyward, are the nude souls awaiting word of their fortune. A taller throng is on the left, where, in prominence, are figures that would—for the sake of the closure of the space of the volume—be akin to Adam and Eve, who are shown in the second woodcut portraying the event of original sin in a lush zoological garden. The smaller figures to the right lend a sense of depth and perspective to the humans who form a vast circle below. The woodcut illustrates two lines from the Bible (only with woodcuts 3, 5, 6, and 7, where otherwise only one is pictured): Romans 14 ("Omnes stabimus ante tribunal domini") and Matthew 24 ("Vigilate, & orate, quia nescitis qua hora venturus sit dominus") that Corrozet translates with emphasis on sight and vigilance, as if to underline the first verb from Matthew:

> Devant le trosne du grand juge
> Chascun de soy compte rendra,
> Pourtant veillez, qu'il ne vous juge,
> Car ne scavez quand il viendra.
>
> [Before the great judge's throne
> A register of everyone will be done,
> Watch out, when before you his word is shown,
> For you know not when he'll come.]

If vigilance is asked of the reader in view of death—"habe mortem prae oculis"—he or she must look closely and watch over the image itself. The image reflects the ocular aspect it asks the reader to share, and it also brings forward, as in the domestic and local scenery of many of the emblems, an uncommon force of detail. The woman who could be from the preceding emblem, now the most prominent figure in this scene, outstretches her right arm toward the terrestrial sphere. She seems to be on the point of holding the world in her hand at the very moment she begs for redemption. The gesture draws the eye to the virtual contact between her palm and the curvature of the globe.

But the hand is not supporting a terraqueous globe as it is known in many armillary spheres, in which the sea and continents are outlined, but a local landscape of three hills dotted with signs of two buildings. A topographic view of what could be a Swiss or Burgundian landscape replaces the totality of the globe in suspension under the feet of Christ. A universal moment is shown in the midst of signs of local places, indeed of a productive agrarian world in which growth overtakes totality.

The Plowman

The shift that operates in this emblem causes the reader's eye to look back upon the landscapes of the previous woodcuts. Domestic or regal interiors in keenly drawn architectural settings prevail. The landscape of Eden (cut 2) is an eerie anthropomorphic garden of delight; the land in which Adam labors alongside a structure (cut 4) is composed of a great outcropping and trees whose branches resemble the tusks of elephants or spikehorns; the land of the pastor and his sheep (cut 12) seems more like a stony surface, a garigue, where the vegetation grows between rocks or is limited to cartographic signs of trees in the distant background. The most striking—and indeed topographic—landscape of all the woodcuts is the thirty-ninth (Figure 26), an image of the plowman, made famous by George Sand and John Ruskin, who furrows a long and spacious field that fills the lower two-thirds of the frame and is set below an unmistakably Swiss landscape beneath a horizon from which radiates the light (or the chimes) from the square tower of a church.[17] Two barns in the medium background have elongated and gently sloping roofs that identify the architectural style of the Bernese overland. The barefoot plowman, dressed in tattered breeches, is old (his face is wizened) but sturdy (his forearm and hand at the plow are in strong contrast to the spindly leg of the skeleton that runs alongside the horses). The courier of death, which wears a scarf-like toga blowing in the breeze and an hourglass that hangs from its neck, puts a whip to the first two of a team of four scrawny workhorses that pull the plow (mounted on an axle at whose ends turn two wheels with radiating spokes). It is hard to distinguish the plowman's hat from a visor drawn around a mass of hair or a coif made of worms of the kind seen in many a *memento mori*. The scene and the landscape meld with the design and mode of production of the emblem. The plowman cuts furrows identical to the lines that a burin incises in the wood of the block. The figure in the cut cannot be entirely separated

In ſudore vultus tui veſceris pane
tuo.

GENE.   I

A la ſueur de ton uiſaige
Tu gaigneras ta pauure uie.
Apres long trauail,& uſaige,
Voicy la Mort qui te conuie.
                        G iij

Figure 26. Hans Holbein, *Les simulachres & histories faces de la mort* (1538). "The
Plowman," fol. Giii r. [Typ 515.38.456] Houghton Library, Harvard University.

from the engraver who follows the lines that must be cut to make the furrows visible.

The superscription arches back to Genesis 1 ("In sudore tui vesceris pane tuo"), the initial book of the Old Testament cited in the first five cuts, but with emphasis now placed on the similarity of this image with that of Adam (cut 3) tilling the soil ("Maledicta terra in opere tuo . . ."), so as to suggest that the world forms a circle or else is conceived as one of many images portraying an eternal return to the earth in a continuum of difference and repetition marked by the alternation of day and night. If so, the plowman in the thirty-eighth cut has traversed the Neolithic revolution by having husbanded horses and, quite possibly, made use of metal in fashioning (in the mimetic register of the image) the hubs and axle of the plow with wrought iron. If, as John Ruskin made patently clear, the image brings its mode of production into the scene, its details command readings that make them *other* than what they seem to be. The rendering of the trees along the edge of the field resembles cartographers' representations on contemporary maps; the four horses follow lines of the field that seem to nudge toward the church in the distance; and the part of the plowman's right heel that is out of the frame implies the presence of a greater space beyond the window of the emblem itself. Most striking, however, is the absolutely minuscule figure of the hourglass, the attribute of the *memento mori* that is placed in all but five of the cuts.[18] Of all the twenty-one woodcuts in which it appears, here the hourglass is least perceptible, but most squarely set at a vanishing point. If Holbein's representation of the revolution of the planets (in "The Astrologer") is kept in view, the relation of the hourglass to a universal "force of attraction" or gravity comes into play.

"Le vice, la mort, la pauvreté, les maladies, sont subjects graves et qui grevent" (818/828) [Vice, death, poverty, illness are grave subjects that aggrieve], wrote Montaigne at the outset of "Sur des vers de Virgile," one of the most "incised" and "engraved" of all of his essays. These "grave" subjects, he writes with sly conviction, cut into our hearts and souls: so too here, and no less literally or with the graphic wit of the *Essais*. The sight line that can be drawn by the grains of sand that fall through the narrow funnel at the juncture of the two spheres—not visible here but clearly implied to be in view—runs downward along the exact axis of the stools that drop from the horse that defecates as it moves ahead. Sand and stool *drop* earthward. The image invites the reader to gloss it with humor enough to find a play of proportion between a grain of sand and a lump of horse manure. It is a visual joke and an invitation to study the relativity of things in the world at large, that is, the relation that objects hold between their size, where they are, and how they move in the rhythms of the visible world. The earth and its place are conjugated in the equally cosmic and comic dimensions of the woodcut.[19]

The gravity and gravitas of the engraving are not limited to the image. They flow into the shape of the typographic design of the quatrain below. The tonic inflection of the capital letters bears visual force: the *A* of "A la sueur de ton visage" [With the sweat of your brow] conveys the compasslike form that Geoffroy Tory and other designers had associated with the majuscule.[20] It occurs in the third line, "Apres long travail . . ." [After long labor . . .] and is inverted in the letter-incipit of the fourth: "Voicy la Mort qui te convie" [Here is Death who beckons you]. *V* (like the Roman numeral heralding Montaigne's "Sur des vers de Virgile") is a monogram or a hieroglyph of the instrument that *engraves* with such uncommon force of gravity, which draws energy and application of force downward, like a hand guiding a burin on a block, into visibility. The arresting sign of the *Simulachres,* Death's hourglass, has as its stenographic analogue the shape of the *V* posed over the *A.* But the *V* of *Voicy* makes the letter visible both before and at the same time as the aural and ocular sense of the deictic "see here" is grasped. The sign of death anticipates or even precludes the effect of its meaning and in that way becomes, in the play of text and image, the pervasive "event" of the *Simulachres.*

It suffices to see how the event suffuses the texts and images at different places. Already in emblem 37, directly to the right of the peddler whom Death pulls by the left arm, and who is accompanied by a mangy mutt at his feet, the scene looks away, toward a wayside shrine on a stunted landscape. The superscription, "Venite ad me qui onorati estis," has as its translation a quatrain whose emblematic sign may be found in the majuscule letters:

Venez, & après moy marchez
Vous qui estes par trop charg[é].
C'est assez suivy les march[é]s.
Vous serez par moy descharg[é].

[Come, and walk after me,
You who are overly burdened;
It's enough to have followed the markets:
You will be unburdened by me.]

V . . . enez . . . *V* . . . ous . . . *V* . . . ous: the majuscule stands in stark contrast to the rounded minuscule that does not convey the visibly cutting qualities of the compass and the style (and stylus) of its form. The extent of its effect and its locational force are shown in the prefatory text under the title of "Diverses Tables" (Figure 27). Affording the cul-de-lampe design that approximates the shape of the *V,* the title of the first line is composed of the majuscule *D* and *T,* in 23-point type, and its minuscule in 14-point. The second line, composed in lowercase majuscule, is set in 12-point. The remaining seven lines are in 8-point minuscule (including the 12-point Roman majuscule to designate Ecclesiastic Doctors and Philosophers). The title is set over the text whose incipit is a historiated *P,* set in a cadre whose stippled background brings forward a floral design on which a bird—whose head and beak arch backward—is perched. The 72-point initial seems to belong to the style of the cartographer and engraver Oronce Finé.[21] No matter what its provenance, the letter disengages those that follow from the text in order—momentarily—to cause the visual design to take precedence over the meaning. *P* inaugurates *Pour,* its remaining three letters, and the *C* of *Chrestiennement* in 10-point Roman majuscule. The uppercases letters of the rest of the page, in the same size, mark the incipit of each sentence and the heading of the substantive that studs the writing: *M,* of *Mort.*

An account of the style, size, and design of the letters and words would seem fastidious were it not for the cul-de-lampe arrangement of the title that continues the downward trajectory to the *V* of (P) *our.* The converging sight lines almost abut the *M* of *Mort* in the second line below, thus causing to disengage from within the very form of the letter *M* the presence of a minuscule-majuscule *v,* the character that encodes the coming—*voicy, venez,* and so on—of Death. From then on a hieroglyph, a product of the very "skeleton" that had been known, at least from Matthias Huss's *La grant dance macabre* (Lyons, 1499) to inhabit printers' studios.[22] A more recent admirer of Holbein, Jacques Lacan (1966, 496), had famously studied, in the "Instance of the Letter in the Unconscious," how the signifier always anticipates the coming of the signified and how it deploys much (and even more) of its meaning in its aural and visual form before it is registered as such. The signifier heralds and acquires a force of fantasy prior to its dissolution into meaning. Thus everywhere in the text the *M* and the *V* would constitute the letters that shift between their functions as graphemes and as locational

Diuerſes Tables de

MORT, NON PAINCTES,
mais extraicẗes de l'eſcripture ſaincẗe,
colorées par Docẗeurs Eccle
ſiaſtiques, & vmbra⸗
gées par Philo⸗
ſophes.

O V R Chreſtiennement parler de
la Mort,ie ne ſcauroys vers qui m'en
mieulx interroguer,qu'enuers celluy
bon S. P O L, qui par tant de Mortz
eſt paruenu a la fin en la gloire de
celluy,qui tant glorieuſemët trium⸗
phant de la Mort,diſoit: O Mort,ie
ſeray ta Mort. Parquoy a ce,que ce
intrepidable Cheualier de la Mort
dicẗ en l'epiſtre aux Theſſaloniques. Ie treuue que là il ap⸗
pelle le mourir vng dormir, & la Mort vng ſommeil. Et
certes mieulx ne la pouuoit il effigier,que de l'accomparer
au dormir. Car comme le ſommeil ne eſtaincẗ l'homme,
mais detiët le corps en repoz pour vng temps,ainſi la Mort
ne perd l'hõme,mais priue ſon corps de ſes mouuementz,&
operatiõs.Et cõme les membres endormiz de rechef excitez
ſe meuuent,viuent,& oeuurent:ainſi noz corps par la puiſ⸗
ſance de Dieu reſuſcitez viuent eternellemët.Nul,certes,ſen

B

Figure 27. Hans Holbein, *Les simulachres & histories faces de la mort* (1538). Incipit of
"Diverses Tables . . . ," fol. B r. [Typ 515.38.456] Houghton Library, Harvard University.

signs that pertain to the space, if not even the landscapes and urban views, of the woodcuts. The plowman might be an instance of the effect; the stout handles he holds as he moves ahead belong to the matrix of the *V.* They are doubled in the image where the dashing skeleton "folds" his leg while putting the whip to the horses. The character of death proliferates so much that it becomes an attribute of space, place, and the graven labors of life and death.

To be sure, the entrepreneurial typographer and poet Gilles Corrozet was aware of what could be done with the play of woodcut and images and textual forms. His *Hécatomgraphie,* as we have seen, draws on what is given in the *Simulachres,* with the difference that the iconography seems more Italianate and of mythological and classical inspiration than the northern style of Holbein. But the process and what it does to create enigmatic events of discourse and space, and to articulate relations between cosmography and topography, are not. In these and other books of emblems is found material to inspire complex designs in which location and locational impulses become driving forces. One of these is the work of Maurice Scève, and another is Pierre de Ronsard, who draws both on the emblem and the design of Scève's emblematic verse. Their topographies are the subject of what follows.

4
A Poet in Relief: Maurice Scève

Maurice Scève has become an industry in French studies. Cursory review of catalogues in every major library reveals the irony that works
on the poet abound but that original editions of his writing are few
and far between. Now and again a library may own a copy of *Délie:
Obiect de plus haulte vertu* (1544), *Saulsaye* (1547), or the royal entry
he prepared for Henry II's visit to his city (1549) for which he was a
principal designer. A later edition of *Délie* (1564) might be found, or
perhaps a copy of his *Microcosme* (1552). After that little comes forward
until the twentieth century, perhaps because Mallarmé's arcane and
precious verse makes it audible or for the reason that psychic intensity
meets an unspoken critical desire, when our planet is in turmoil and
decay, to venture into other worlds. The task of this chapter is not to
escape into those worlds but, more modestly, to gather a sense of some
of Scève's coordinates and coordination: of space and place through
the heritage of the emblem; of the tension felt between cosmography
and geography; of the adventure of alterity in the hills and valleys of
consciousness when the verse moves in and about the space and form
of its own idiom; in the presence of mapping and self-locating images
that serve to determine the poet's self-estrangement and, as a result, in
the implications of the poetry for what concerns topography. The task
entails a partial reading of *Délie* and a view cast upon the local setting
of *Saulsaye.* It requires appreciating each as a composite work crafted by

the poet and, to a marked degree, by the artists of emblems, city views, and landscapes. One of the latter, it will be argued in passages concerning *Saulsaye,* is Bernard Salomon, the gifted draftsman and designer with whom Scève frequently collaborated during his years in Lyons before and around 1550.

Délie

The locational bent of *Délie* is obvious in its play of the 449 square-like dizains and fifty emblems in concert with them. Known as the first French *canzoniere* in the tradition of Petrarch, and affiliated with the intellectual ferment of Lyons, the rich commercial center between Paris and Italy in the early and middle years of the sixteenth century, the poem can be seen as an arcane geography of subjectivity, a mental map of love and speculation on the nature of place: on what it means, in a vital relation with the unknown, to sense not to be where one is wherever one happens to be.[1] As a rule it can be said that the poems embody the effects of self-displacement within a geography the poet invents with words, emblems, and images. For the late Gérard Defaux, author of a rich and passionately crafted critical edition (2004) of *Délie,* Scève is a poet of desire and drive. A Christian Prometheus, he dares to steal fire from God and to wrestle with a force of love both carnal and devout, indeed of "Marial" quality (related to the Virgin Mary as she was seen in the legacy of iconography reaching back to the fifteenth century). An artist living in a postlapsarian condition, Scève dares to commit himself to and to take pride in the sin of idolatry. Idolizing *Délie,* not just the anagram of *Idée* (Idea, the abstraction of an idea or a principle of love), he writes a masterwork of continuous passion for a woman, in flesh and blood, willfully confused with the mother of Christ and the figure of Laura, born in Cabrières, the beloved mistress of Petrarch (1304–74). Scève's existential relation with the Vaucluse had even inspired him to seek and indeed (so the story goes) discover Laura's tomb. *Délie* is composed in the midst of the Petrarchan revival in Lyons, and it is a reflection both of the political conditions of the time and of the character—the *genius loci*—in which the work is written and published.[2] In all events the act of writing frees him—*le délie*—from the contingent time and space in which he lives and of which the poem often writes.

Délie sets into an order of difference and repetition an arcane sum of dizains (449) with a round number (fifty) of emblems. At times the dizain that follows or is found under its emblem recoups, with ever-so-slight variation, the words of the motto that *turn about* the image in the visual field, adjacent to the inner border of the ornate cartouche in which the image and text are contained. The style of reading that the poem requires is cinematic (Alduy 2003, 34–35). The eye moves to and from the adjacent text and its woodcut image, and from the inside of the cartouche to its decorative matter, and from there, a decorative edge, onto the visual aspect of the dizain as the eye beholds the image and the ambient lines of verse. And each square block of verse "can be read in its punctual density, as a *unit of textual space*" (Risset 1971, 95) in which anagrammar, by which the eye finds letters of words embedded in the gist of others, "disperses and brings together, opposes and puts into play its own phonemes in a concentrated interaction taken up and dispersed through other dizains in the totality of the work" (ibid.). The textual geometry and architecture find analogues in the decorative schemes, and in their difference and repetition as emblems punctuating the work, they engage multiple ocular and cognitive trajectories or itineraries. The latter can take place simultaneously, in at once conscious and unconscious registers. The narrative of the artist's and lover's travails, moving tortuously toward positive resolution in the ordinate direction of the text, from one dizain site to the next, is countered by an itinerary of self-destruction in the concatenation of the emblems (Charpentier 1984, 41). As of the title page the book suggests a topography of travel: an image of a rocky island, recalling for many readers the vicinity of a Mediterranean archipelago, is seen from the four corners of the page by as many wind heads that blow currents of breeze and tempest from the four cardinal points (Figure 28). The poems and the visual matter are composed to move multifariously.

The cinematic character of the work owes to the way it can be read backward and forward at once, and to how the emblems force the eye to take cognizance of the way it gazes on and reads—haptically—the poems and images. The first five ten-line poems are punctuated by the initial emblem that gives way to groups of eight between each of the following fifty emblems, at the end of which the 447th dizain counts among one of three comprising the end of the poem. The rhythm of grouping in sets of eight yields an effect of a "spiritual automaton" or poetic and projective machine, even an experimental film whose montage and placement are designed according to a recurring cadence of images and units of writing.[3] By every standard it is categorically impossible

DELIE.

OBIECT DE
PLVS HAVLTE
VERTV.

ADVERSIS

DVRO.

A LYON
Chez Sulpice Sabon, pour Antoine Constantin.
1 5 4 4.
Auec priuilege pour six Ans.

Figure 28. Maurice Scève, *Délie: Obiect de plus haulte vertu* (1544), title page. [FC5. Sce925.544d] Houghton Library, Harvard University.

for readers of the twentieth and twenty-first centuries *not* to engage the poem through the experience of the seventh art.

The sense of seeing and touching images moving throughout *Délie* tells much about its topographic construction and locational drive. The sheer difficulty and obscurity of the language of the poem notwithstanding, the most immediately striking visual forms are the decorative surrounds that frame the emblems. Each emblem is composed of an ornate cartouche whose strapwork curls about the frame as if into a two-dimensional space; or backward, along a receding path to evince the same effect of volume of a form suspended in an area of blankness; sometimes the frames have circles cutting through them and are shaded on one side to imply light cast upon a thick body of metal or leather with which they are crafted or even grommetted; now and again the corners bend back to shape consoles with snail-like spirals that support grotesque figures—sometimes of simian appearance, at other of svelte and supple bodies drawn in the Fontainebleau style—that inhabit or support them. In every instance they are mirrored designs that repeat, with some variation, on each side (left or right and sometimes top and bottom) the shape of the others. At the upper midpoint of many of them a mascaron stares forward frontally, directly at the reader, as if to beg wonder, fear, admiration, shock, or even laughter. A good number display heads of chimerical fauna—goats, canines, monkeys—that hold in their mouths the ends of festoons or swags of flowers and fruits swinging and curling down and about the design. They seem not only to enhance the effect of enigma but also to look at the reader whose gaze strives to wander about and to decipher the poem.

The detail and intricacy of their form exceed those of the images held within. The pictures have specific iconographic reference, often to classical mythology, and they also, like the subjects in Guillaume de La Perrière's *Le théâtre des bons engins* of Parisian origin (published by Denys Janot, circa 1539), often represent (or moralize) the labors of everyday life. In comparison with the work affiliated with Corrozet the images are crude and of far less visual force than the frames that contain them. The unabashed admirer of *Délie* may tend not to find the images sublime because the language of the poem turns them into "dissimilar signs" that, in their inadequacy or incongruity, "appeal to the immediate transgression of sentience, and exist only in order to allow us better to understand what are the most grotesque forms of representation, the most material and lowest as well, that succeed still in the best ways of expressing the highest spiritual realities."[4] The poverty of the images in the surrounds is redeemed by the nature of the poem tied to the printed matter (in small Roman majuscules) that turns about the image and that is a buffer, a *passe-partout,* or a parergon between the picture and the cartouche.

Epigrams and Emblems

A historical bias on the images is informative. If, as in the early 1540s, Parisian printers owned abundant stocks of woodcuts—of the kind seen in La Perrière's *Théâtre des bons engins* or in the *Hécatomgraphie*—it may be that they came to Lyons either little or late. Denys de Harsy publishes his own editions of Alciati's emblems, La Perrière (ca. 1539), and Corrozet in pirated form and without illustrations. The *Délie* might have been the very first book of emblems in Lyons, yet as a book that not only captures the spirit and method of the genre but also mobilizes it with new and unforeseen effects.[5] That "La teneur du privilege" in the 1544 edition states that Antoine Constantin, "marchant libraire demourant à Lyon," is allowed to "print or have printed by such printers of the cities of Paris, Lyons, and others as he wishes, this present book treating *[traictant]* of Loves, titled DÉLIE, with or without Emblems, during the time and end of the six forthcoming years" (Defaux 2004: 1, 2) can have multiple implications: the reader may already know the emblems and their devices in his or her mind's eye or memory. Perhaps the editor is, as many an emblem has shown, "between the mill and the fireplace" or at odds about what can be obtained for the edition. Constantin may have launched the book before all of the emblems were finished or set in place, as he awaited their completion, and as a result the hieroglyphic character of *Délie* would reside at once in the verbal and graphic form both of the poem and the intellect (with emphasis on memory) the reader brings to them. With emblems an image, however approximate it may be, suffuses the ambient writing and brings it at once—as it was suggested in Rabelais—to concomitantly higher and lower meanings, respectively, to the realm of the ideal (the ideogram) and the brute impact of physical reality (the rebus) (Céard and Margolin 1984, 1:65, 284–85). Already, in the dedication, "A sa Délie," the poet states, seemingly unequivocally, in an elegantly self-deprecating *captatio,*

Je sçay asses, que tu y pourras lire
Mainte erreur, mesme en si durs Epygrammes: (2004: 2, 3, lines 5–6)

"I know well that here you can read many errors, even in such difficult epigrams": from the standpoint of a higher—and highly geographical—sense, the many errors would be tied to the poet's errant ways in his imitation of Petrarch, the existential poet who wrote his *Canzoniere* on the heels of his sins and his interrupted travels to and from France and Italy.[6] This inflection would chime with the first line of the poem: "L'oeil trop ardent en mes jeunes erreurs / Girouettoit, maul cault, à l'impourveue" (D1, lines 1–2). [This overly ardent eye in my youthful errancy / Spun like a weathervane, imprudently, unexpectedly.] And from that of a lower, indeed highly localized and topographic sense, the errors the reader is invited to discern can be those of a typesetter, such as the construction that contains the "universe" (sun and moon in shadows, concentric circles of asterisk stars, and two crescents mirroring each other) shown upside down in the edition of 2004 (between D14 and D15).[7] The same double inflection would hold for the graphic wealth of *Epygrammes*. Not *epigrammes,* but *Epygrammes:* the orthography suggests that they belong to the world of emblems, following Alciati's celebrated definition of the character of originary (hence both aural and visual) writing, *epigrammata sive emblemata,* which refers indirectly to Egyptian hieroglyphs because of the implied presence of Egyp(t) in the letters of the word. Far-fetched as it may seem, the reading reaches into the geography of *Délie* when, in one of the most decisive of all of the dizains, Egypt figures at a cornerstone of the poem in which in his longing after the departure of his beloved the poet chooses a timorous hare to be his totem:

> Car dès le poinct, que partie tu fus,
> Comme le Lievre accroppy en son giste,
> Je tendz l'oreille, oyant un bruyt confus,
> Tout esperdu aux tenebres d'Egypte. (D129, lines 7–10)

> [For as of this point, you having departed,
> Like the hare crouching in its lair,
> I extend my ears, hearing a confused noise,
> Totally lost in the shadows of Egypt.]

An extraordinary counterpart is made between the hare's den and the image of Egypt. A space of frightened retreat is confused with a toponym of a fabled place at the origin of the hieroglyph. The totemic emblem of fear and of aural intensity is enhanced by the shape of extended ears listening to the faint sounds reverberating from one time and place to another.[8]

Whichever way the reader chooses to go, left or right or about and around, like the weathervane, the metrical and the subjective dimensions of *erreur* and *Epygramme* point to a complex geography whose points of coordination can be found in the emblems and their recurrences; in the cartouches and their placement and variation; in the toponyms scattered about the dizains; in the emblematic and figural dimensions of the words and letters and their sites and situations within the gridding of many characters in the maplike form of the poems. Here the plan of variation in and of the cartouches and their geometrical relation with the *gîte*—and the gist—of the surrounding dizains are pertinent. Five forms constitute the frames for the virtual "epigrams": a rectangle, a circle, a lozenge, an oval whose axis is horizontal, and a triangle. One oval, whose double axis is aligned vertically, recurs eight times. None of the same forms is immediately adjacent to any of the others, but in sum a quasi-even distribution of the shapes is assured. The three rectangles occur once each (totaling nine times); the circle likewise; the lozenge and its three variants occur eight times, as does the triangle. Only the vertical oval, of a single design, occurs eight times without any variation in form.[9] They are at a distance from one another of fifteen or twenty dizains and thus assure at once an unpredictable order and a harmony. Where one falls in the text its recurrence elsewhere can be paratactically linked to it and to what its adjacent poems are expressing—as it equally may not. The enigmatic semaphores seem to be akin to navigational flags that bring the reader to compare *where* the writer-lover finds himself in his turmoil in different times and spaces of the poem.

A Spider's Eye

One of the most geographical of the cartouches and emblems (Figure 29) is set above the 420th dizain.[10] A spider (or a black smudge) is adjacent to the axis of its circular web and filaments. The frame is one of three lozenge patterns that come late in the volume, first encountered in emblem 37 ("the viper that kills itself"), whose motto is "Pour te donner vie je me donne mort" (To give you life I give myself death), between dizains 239 and 240, and emblem 40 ("the rooster that burns itself") within the frame of "Plus j'estains plus l'allume" (The more I extinguish, the more I set it ablaze), located between dizains 346 and 347. On cursory glance the design is an ocular enigma. The web is shown scenographically, as it might hang from a branch or a rafter of the kind seen in Holbein's shelter (Figure 24). Yet the lozenge can be seen as an

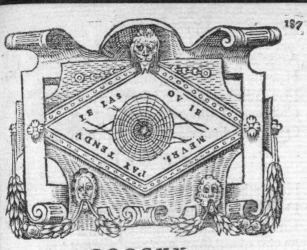

CCCCXX.

Au doulx rouer de ſes chaſtes regardz
Toute doulceur penetramment ſe fiche
Iuſqu'au ſecret,ou mes ſentementz ars
Le plus du temps laiſſent ma vie en friche,
Ou du plaiſir ſur tout aultre bien riche
Elle m'allege interieurement:
Et en ce mien heureux meilleurement
Ie m'en voys tout en eſprit eſperdu.
    Dont,maulgre moy,trop vouluntairement
Ie me meurs pris es rhetz,que i'ay tendu.

CCCCXXI.

Mont coſtoyant le Fleuue,& la Cite,
Perdant ma veue en longue proſpectiue,
Combien m'as tu,mais combien incite
A viure en toy vie contemplatiue?
Ou toutesfoys mon cœur par œuure actiue
                          Auec

Figure 29. Maurice Scève, *Délie: Obiect de plus haulte vertu* (1544), spider emblem over dizain 420. [FC5.Sce925.544d] Houghton Library, Harvard University.

anamorphic square, whose point of view is that of a bird's eye, from a three-quarter perspective. To these two angles on the image a third can be added, that of an ichnographic point of view above or below the object portrayed: for, as the 419th dizain indicates, the lover-writer's idol is so sublimely beautiful that, without committing the sin of perjury, he can avow that her aura *remplit l'oeil* (line 3) (fills his eye) and goes to the quick of his heart, lighting a new desire that stirs *tous les sens* (all his senses) and goes to his head. Expressed here and elsewhere, the Petrarchan formula of the eye whose bright shafts of light shot from the pupil to dazzle the lover is obvious, even hackneyed: and so possibly also the remark, in the last quatrain, that his idol would even move "this great Roman censor" who, by allusion, is the stern and immutable Portius Cato. Endless emphasis on the eye causes the cobweb to resemble a sort of skeletal pupil, an eye whose former radiance and arrows of light are now the dark lines of an inked engraving. The cyclopean spider-as-eye, in contrast to the binocular views on the part of the three mascarons that stare at the reader, would confirm, as other dizains related to the specific context, that the eye is overexposed, burned, and even immolated. The more it burns, the more it extinguishes, and vice versa.

But the arrangement of the motto, its first clause, "J'ay tendu" (I've tendered), read upside down on the lower left edge, then clockwise, right side up from the upper left, "le las" (the web), then from the lower right, "ou je" (where I), and finally upside down again from the right, "meurs" (am dying), forces the reader to turn the book around and to gaze upon it, like a map, from all cardinal directions. As the book turns, so does the eye upon its axis. All of a sudden it is obvious that the cobweb is drawn in exactly the same fashion as Pieter Apian's polar projection of the earth in his *Cosmographie* (Figure 14). In the design of the image we witness a hemisphere in a web, and a finely crafted set of orthogonal lines of longitude and concurrently circular lines of latitude. "I've tendered the web where I die." The spider-poet's strategy is to put himself at the axis or the origin of a plan of an *imago mundi* whose oceans and continents are invisible or not yet set in place. The difference and identity of a world space and of a site of attraction could not be more clearly drawn in the same "epygramme." Seen thus the emblem broadens the scope of its interpretation and its relation to its provenance. The figure in which the poet is caught in the tender trap of his own design, as Defaux notes, "has nothing original about it"

(2004, 2:441). It is found in Denys Janot's edition of La Perrière (1539), and it could be inspired by Petrarch's *opra d'aragna*. Defaux notes that Scève "probably remembered this sonnet" and "had the Psalms [9, 16] in mind when he composed the device" (ibid.). Including the iconography of Pallas and Arachne, the wealth of conventional associations is less important than what they do in constructing their topography. The *las,* the web, with the inflection of *in laqueo* in the Psalms (from *laqueus,* a noose, a snare, or trap), as in Ovid, *in laqueus cadere,* might not be far from *laqueo,* "to adorn with a paneled ceiling" (such as Tory's trellis seen in Figure 7), or to design a reticulated form, a net of the kind used to catch errant prey. The net would be the entire form of the *Délie,* the very webbing of its lines and points of which this emblem and its adjacent poems are sites and coordinate crossovers. The device is far less suicidal or doubly bound than one of a tactical ruse: the poet offers the image of dying—of finding his sepulture and his resting place—in the space of his own creation in order to draw the reader's eye into its trappings.[11]

The poet carefully maps his own demise. Which is what dizain 420, closely correlated in form with the image, suggests: the poem moves from a trap, tendered to the reader in the first line, that begs association of the circular image with a "soft" wheel of the beloved's eye to that of its "hard" counterpart, the wheel of torture (or possibly, of ill fortune). The *doulx rouer* of the idol's "chaste" gaze blends into "Toute doulceur" [all sweetness] that is struck at the secret point—at the bull's eye—of the image, now seen as a target as well as the beloved's pupil, a spider web, and a conic projection—where his burning passions mostly "leave my life in abandon" (*en friche,* in the fourth line, which bears strong erotic innuendo). She lightens *(allege)* his pleasure (or pain) over all other sentiments. And in this *mien* (line 7, what is mine, what belongs to me, but also this "mien"), the poet goes off and away, in a lost frame of mind. Wherever he goes he errs, but along the way he makes a map of his meanderings by the fact—and the homonym—that he sees himself in a lost bent of mind. "Je m'en voys tout en esprit esperdu": to write clearly and distinctly that "I go off" and by implication "see myself" lost means that the "I" is not lost; rather, it finds its psychic and geographical bearings in its own statement and the printed image of that very statement. The couplet that serves as a finale binds the voice of the poet to the target-eye-web-map when it reprises and varies the motto. In stating that he "too voluntarily" (line 9) dies when taken in the *rhetz* or net that he has spun and tendered, the act of will, however excessive it is taken to be, is an affirmation and a commitment to the project of crafting a world through creative idolatry.

The cosmography or total writing of the sphere—and the hope it engenders—comes forward from the topographic setting of dizain 412 that follows, in which a city view, of a local perspective, seems to be a fitting complement to what precedes:

> Mont costoyant le Fleuve, & la Cité
> Perdant ma veue en longue prospective,
> Combien m'as tu, mais combien incité
> A vivre en toy vie contemplative?
> Ou toutesfoys mon coeur par oeuvre active
> Avec les yeulx leve au Ciel la pensée
> Hors de soucy d'ire, et dueil dispensée
> Pour admirer la paix, qui me tesmoigne
> Celle vertu lassus recompensée
> Qui du Vulgaire, au moins ce peu, m'esloigne.

> [Mountain by the shore of the River, and the City
> Losing my sight in a long perspective,
> How much have you, but how much incited
> To live in you a contemplative life?
> Where always my heart, in active labor,
> With my eyes, rises to the sky the thought
> Beyond worry, ire, and grief dispensed
> To admire peace, that proves to me
> This virtue alone rewarded
> Which, at least removes me from the Common.]

The poem arches back to others that begin as topographic views of Lyons (especially dizain 17). The vantage point is between that of a bird's-eye view on the city, as if from Bernard Salomon's contemporary view of Lyons (Figure 30), where Mount Fourvière stands above the confluence of the rivers Rhône and Saône. The poem drifts from a "site-specific" place to a locus of abstraction, to the topic of the solitary and contemplative life, a life where the poet would neither desire to desire nor desire not to desire. The *active labor* that is taken to be the poem as its weaves its webbing stands in contrast to the icon of Ptolemy, the cosmographer known always to look skyward in his contemplation of the mechanism of the planets. Here the image is shifted into a philosophical register, but without loss of the geographical bearings, cited in the first two lines, in which contemplation takes place.

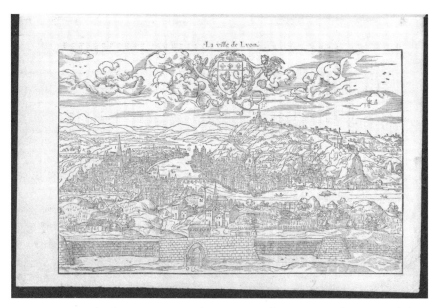

Figure 30. Bernard Salomon, city view of Lyons, in Guillaume Guéroult, *Epitome de la corographie d'Europe, illustré des pourtraits des villes plus renommées d'icelles* (Lyons: Balthazar Arnoullet, 1553). [Typ 515.53.439 F] Houghton Library, Harvard University.

If Scève's poems are "durs Epygrammes," and if their discourse is riddled with hieroglyphic signs—of the kind that Dainville (1964/2002) and Delano-Smith (2007) call a geographical idiom—the shape of letters cannot fail to have iconic form. In the space of the dizain a turn, up and away from the "tribe" of common mortals, from the Common or Vulgar, would mean more than it says. For Geoffroy Tory the majuscule *A* had been designed in the manner of the *V* and was allegorized to include its mode of creation within its own form, as a pair of dividers (1529, fol. xxx v), possibly as a model for the broken analogue that Corrozet had shown in his *Hécatomgraphie* (Figure 20). For Holbein the *V* in the *Simulachres* is the stylus that engraves the sign of death in *Mort* and other words in which orthogonal lines converge. For Montaigne it is a monogram that both marks and becomes a measuring instrument and a form that conflates sexual difference when seen in two ways at once (Conley 2002). For Scève, similarly, it is all that and a perspective that opens on the world at whose nadir is the origin of sight (at the site where it indicates the line on which it is placed), an *angle* on the world.

In *Vulgaire,* in which the choice of the majuscule is capital (even in the economic sense), the people designated by the word might be those looking from below up to the poet where he stands on the *Mons Veneris* over and behind the city. The virtue of the poet awaiting from above *(lassus)* can be projected to be drawn, as if from the network or grid of the map seen supra, in the preceding poem and emblem.

The itinerary taken to soar over the city, to witness it from above, and to contemplate greater forms in an *alterior sensus* mythifies or even invents the space in which the adulation of Délie takes place. He turns Lyons into an object of poetry invested in the virtues of its name. The augmentation of "place-value" through affiliation with the site of origin of the poem has much to do with the Promethean character of the implied author. In the celebrated dizain 77 he puts himself in the role of the hero having competed with the gods, the audacious and sly Titan who, according to Ovid and Hesiod, fashioned the human species.[12] The poem is of one sentence, one breath, in which resolution comes only with hope:

Au Caucasus de mon souffrir lyé
Dedans l'Enfer de ma peine eternelle,
Ce grand desir de mon bien oblyé,
Comme l'Aultour de ma mort immortelle,
Ronge l'esprit par une fureur telle,
Que consommé d'un si ardent poursuyvre,
Espoir le fait, non pour mon bien, revivre:
Mais pour au mal renaistre incessamment,
Affin qu'en moy ce mien malheureux vivre
Prometheus tourmente innocemment.

[On the Caucasus of my suffering, tied
Within the Hell of my eternal pain,
This great desire of my forgotten good,
Like the vulture of my immortal death
Eats at the mind with such a fury,
That consumed by so ardent a pursuit,
Hope, not for my good, makes it relive:
But for misery to be reborn incessantly,
So that in myself in this my unfortunate life
Prometheus torments innocently.]

The dizain suggests that the "I"—tied to "the Caucasus of my suffering"— is part of a psychomachia in which physical and spiritual love are in conflict and that, as high as he may be, the poet suffers in making the world of

his poem and of his name. As Defaux states with passion indeed rivaling
that of his poet,

> In sum everything indicates that both here and elsewhere Scève looks skyward,
> that he evolves in altitude, and that he is eager to tell us that his "great de-
> sire," an always living, tirelessly *(inlassablement)* active desire, is not only a desire
> for the "good" in general . . . , but also and especially a desire of "*his* obliged
> good" [Defaux reads *oblyé* in the juridical and moral sense, the Latin *ob-ligo* and
> *ob-ligatus*], that is, first of a desire for personal wealth—"my good"—and then a
> good toward which he sees himself "obliged" and to which, for having probably
> staked all his faith, all his force, he is forever "ob-lyé." (2004: 1, xciii)

Defaux introduces a hyphen to mark how the prefix *ob-* can imply linkage toward, against, upon, over, and down.

He adds that this poem is one in which "all its words are images.
We no longer read, we see" (2004: 2, 114). The hieroglyphs that are the
letters, words, and spacings of the dizain suggest as much and more.
In the first line the infinitive substantive, *mon souffrir,* beckons recall
of the poet's own device, *souffrir non souffrir,* which become the final
words of *Délie* and *Saulsaye,* implying that the poet suffers in the creation of a signature to which he is Prometheus bound. He suffers because
his name is that of the mighty other for whom he lives in deficit and
infinite admiration. The mention of the Caucasus Mountains, in which
altitude is countered by all that falls in the toponym, from *casus,* designates limitrophe regions not far from Scythia that would be at the line
of geographical alterity. The poet is at the limit of his own obligations,
"like the vulture *(aultour)* of my immortal death." *Aultour* can resound,
like the first line, with signs of the self to which the bird is compared:
to the *autheur,* the other "around whom" the bird that *is* the author
turns in the sky above. Capitalized, it brings the reader to study it as a
hieroglyph. *Aultour* inaugurates its path of flight with *A,* the sign of the
surveyors' dividers and the icon of the origin both of writing and of
visibility. Where the *V* of *Vulgaire* opens a line of sight from below, the
A of *Aultour* spots it from above. The resulting point of view is seen as
a calligram in the majuscule of its name. It circles above and around—
aultour—the body of the poet it eternally vivisects, and by cause of the
myth of Prometheus, like the "auteur," it is haplessly obliged forever to
pick and tear in order to assure the victim's "immortal death."

The hieroglyphic letter coordinates local and cosmic space. Hope,
the figure that leads the poem to its resolution in the last three lines,
bears the idea of a sphere, a memory image of the greater world whose

recall (as shown in chapters 1 and 3) rescues the poet (in the narrative of the dizain) from his consummation and incineration. In the last three lines the vulture is suggestively turned into a phoenix—*au mal renaistre incessamment* (in pain to be endlessly reborn) in the scene of suffering atop the mountain. The mythic space in which the writer is transcendently immanent (or immanently transcendent) is both identical to and far from the top of the Mount Fourvière mentioned in nine other dizains. Nor is it remote from the ninth emblem, *La targue* (the shield) immediately below (Figure 31). Set in the second of the three lozenges (and in its own first appearance in the poem), the shield bears some resemblance to the target that inaugurated the *Hécatomgraphie* (Figure 22). Pierced by an arrow, its central fold forms an illusion of three-dimensional space. It is a "dissimilar sign" (Miernowski 1997) in that it refers to the iconography of the combat of love and war and to the configuration of the image in which it is placed. The device, "Ma fer- / met[é] / me / nuict" (My obstinacy does me harm), like that of the fifty-sixth emblem, can be read from four sides. The shadowed hatching on the bottom and in the two concavities of the shield causes *nuict* to be seen and heard as a play of light and dark or of diurnal and nocturnal reason. The two horizontal angles of the lozenge that open from the left and right sides bear resemblance to the *A* of the dizain above, particularly because the lion and canine on either side have sight lines confirmed by the geometry of the lozenge. If, in the debates over the anteriority of the text or the emblem or the supposition that Scève had the illustrations designed for the book, the gist of the first quatrain of the subscriptive dizain suggests that they are coeval. The poet avows that he is complicit in this "sweet" *(doulce)* battle that without resolution holds him in suspension. "If I am stricken from one side the other cuts [*taille*, as an engraver incises an image], / all *[tout]* on a level with what upholds me." The words can refer to the struggle of love and to the image and its placement in the frame of the poem in the mode of Petrarch's *Canzoniere*. The one and the other, like the two mascarons, look from either side upon the evidence of a metaphorical wounding of the author. In the middle quatrain the poet avows that the one side maintains "that hope *[espoir]* is no more than a vain umbrage," a shadow, while the other states that desire is a rage. The first leads us "under a blinded night *[nuict]*" (line 8). Hope, initially figured as a globe, is now enshrouded in shadow. It is properly a *sphere* conveying Pauline ideology, which continues to lay stress on the poem's production of space that is both *hic et nunc,* on the printed page, in and about the side of the

LXXVIII.

Ie me complais en ſi doulce bataille,
Qui ſans reſouldre,en ſuſpend m'entretient.
Si l'vn me point d'vn coſte',l'autre taille
Tout rez a rez de ce,qui me souſtient.
    L'vn de ſa part treſobſtine'maintient,
Que l'eſpoir n'eſt,ſinon vn vain vmbrage:
Et l'aultre dit deſir eſtre vne rage,
Qui nous conduit ſoubz aueuglée nuict.
    Mais de ſi grand,& perilleux naufrage
Ma fermete' retient ce,qui me nuict.

LXXIX.

L'Aulbe eſtaingnoit Eſtoilles a foiſon,
Tirant le iour des regions infimes,
Quand Apollo montant ſur l'Oriſon
Des montz cornuz doroit les haultes cymes.
Lors du profond des tenebreux Abyſmes,

Figure 31. Maurice Scève, *Délie: Obiect de plus haulte vertu* (1544), emblem of the *targue* over dizain 78. [FC5.Sce925.544d] Houghton Library, Harvard University.

production of the book, and in a cosmic realm by which the two forms and local sites are compassed.

In its entirety the poem becomes a mobile topography. The most dramatically spatialized of all the dizains, and perhaps the most remarkable of the collection, would belong to what geographers call a potamography, a description of rivers. Before the middle of the century woodcut topographical maps of Lyons draw the eye to the confluence of the rivers Saône and Rhône. The "road map" of our age, in which broad colored lines draw across the land on a scale far greater than their background, has as its analogue the river map of the sixteenth century. The river, whose curved and tortuous lines often resemble tendrils drawn through the spaces in which they are represented, is generally a sign of circulation, movement, and commerce. Inland cities are valorized according to the rivers they adjoin. The fantasy of movement, of the generation of force and affection, seems to follow the path of rivers. In the famous edition of Petrarch that Alessandro Vellutello edited in Venice (1514) and that no doubt informed the composition of *Il Petrarca,* published in Lyons in 1544, one of the first great topographical maps is appended to the text in a fashion anticipating a *carte du tendre,* an affective map, with the exception that no distinction is made between allegory and description.[13] A paratactic map of the Vaucluse and its environs (paratactic because different events in the lives and love of Petrarch and Laura can be located simultaneously in different places), it draws the swirling swath of the Durance from the poet's retreat, at the Fontaine de Vaucluse, to Carpentras, Sisteron, Cabrières (the birthplace of Laura), Avignon, and the mighty Rhône. A topographical map far more detailed than Oronce Finé's woodcut and manuscript versions of the same region (1551, fol. 54r), the map is both a clue to the nature of the region and an invitation to the reader to coordinate the signal events in Petrarch's life and love with the sites denoted in the woodcut.

A first great literary map, Vellutello's projection may also be what adumbrates the description of the rivers Rhône and Saône in the seventeenth dizain, a maplike poem with which many readers gain access to the otherwise closed character of *Délie* and its landscapes:

> Plus tost seront Rhosne, et Saone desjoinctz,
> Que d'avec toy mon coeur se desassemble:
> Plus tost seront l'un, & l'aultre Mont joinctz,
> Qu'avecques nous aulcun discord s'assemble:
> Plus tost verrons & toy, et moy ensemble

Le Rhosne aller contremont lentement,
Saone monter tresviolentement,
Que ce mien feu, tant soit peu, diminue,
Ny que ma foy descroisse aulcunement.
Car ferme amour sans eulx est plus, que nue.

[No sooner than Rhône and Saône are disjoined
Will my heart be undone from you:
No sooner will the one and the other Mountain be joined
Than will discord gather between us:
No sooner shall we see you and me together
Than the Rhône meander slowly upstream,
The Saône mount impetuously,
Than this passion of mine might ever diminish,
Nor that my faith ever even falter.
For without them firm love is only bare.]

The dizain is an affective projection in which distortions and aberrations of named rivers turn the poem into a map of intense passion. The inversion of the force and flow of the Rhône and the Saône can be glossed topically (and typically) by locating the "source" from which the poem springs in the classical and medieval canon.[14] But it can also, even if tenuously, be traced in the graphic design of *Délie*, where it both assembles and disassembles its own space in a magnificent flow of nine lines that come to a stop before the tenth and clinching decasyllable shoots as if from a rapids surging down a deep valley, what has its hieroglyphic correlative in the majuscule that marks the *Mont* of the third line, and not the "contremont" of line 6. In the thirty-third emblem of Corrozet's *Hécatomgraphie*, in praise of fraternal friendship, a wise and wizened father tells his sons that their lives ought to be led to "avoir paix, & amytié ensemble" [to share peace and friendship together] and "que si aulcun de vous se desassemble" [if any of you break away] from the bond, you will immediately see loss befall you. The breakage that does not break, indeed the unruptured rupture of Scève's poem, is captured here in its elegiac tone and its anticipation of immutable constancy—no sooner than . . . than . . . than . . . will my faith ever diminish—that goes with the flow of the sentence. The past master of the Petrarchan oxymoron bends new shape into the French idiom. The rivers will not be disjoined until "from with" *(d'avec)* you my heart may splinter; before this departure "from with" each other no "discord will assemble" so as to rhyme with *ensemble; feu,* the inner blaze of the poet's passion, rhymes graphically with *peu* but does not diminish as the tonic verb would indicate at the end of the line. But these melding contraries are tributary to the geographical oxymoron at the juncture of the rapid and violent flow of the Rhône whose waters descend from the Alps, fill Lac Léman and irrigate the valleys of the High Savoy before meeting the smooth and slow meander of the Saône down from the rolling hills of Burgundy. The rivers are joined in the first line and thus set the point of view of the speaker squarely in Lyons, at the foot of Mount Fourvière, at a site whose fantasy of inversion is memorable. The poem infuses a common place with a mythic force of love. The seeming intensity of the verse that owes to an absence of psychic relief in the heights and depths or imaginary orography of the poem is tied to the origin, to Lyons, for which the lyrics are an indirect advertisement.

The mercantile dimension of the verse hardly diminishes its own force. To the contrary, its materiality, indeed its exchange value, relieves it of what might be called—if the poem is read as a narrative, and not as a complex webbing of signs of infernal passion and suffering—amorous congestion. In dizain 208 the "flourishing" Rhône, its golden sands and argentine waters bordered by cities and castles, and on which vessels carry goods (no doubt thanks to the teams of horses and men who guide them upstream and down), is comparable to "his lady," but with erotic innuendo that makes desire exceed the motivation for profit. He watches the Rhône swell and "enter into the Sea" (line 10) as if what he saw were carrying his fantasy of fulfilling Délie.[15] The mix of mercantile and erotic registers in the allusions to the two rivers is found in the later dizains, after the seventeenth and twenty-sixth poems, in which the greater topography is established. After he sees himself portrayed in the face of Mount Fourvière, the *mons veneris* that his brushes have depicted—"Je voy en moy estre ce Mont Forviere / En mainte part pincé de mes pinceaux" (D26, lines 1–2) [I see myself being their Mount Fourvière in many places painted with my brushes], the correlation with Apian's distinction of cosmography and geography (see Figures 2 and 3) could not be clearer. It is eroticized in the poem in which the rivers at the foot of the mountain meet, mix, and marry, from where they begin their travels southward, through the valley of the Vaucluse that had been dear to Petrarch, to the sea where they eventually die:

> N'apperçoy tu de l'Occident le Rhosne
> Se destourner, & vers Midy courir,
> Pour seulement se conjoindre à sa Saone
> Jusqu'à leur Mer, ou tous deux vont mourir? (D346, lines 7–10)

[From the West don't you notice the Rhône
Turn away and flow to the south
Only to be conjoined with the Saône
As far as their Sea where the two will die?]

In geographical writing it happens that at the site where the rivers flow
into each other, they lose their names.[16] The poem adds such elegiac and
emotive force to the commonplace that the reader can easily imagine
the speaker of the dizain gazing mutely at the waters and seeing in the
abstraction of their flow at once a lyrical drift, a mirror of constancy,
and tremors of disquiet over the impending loss of the image—Délie—
that has become the very landscape through which the river flows.

From *Délie* to *Saulsaye*

The river has unparalleled graphic presence less in *Délie,* where geography and fantasy abound, than in the *Saulsaye,* the eclogue that the
enterprising editor and polymath Jean de Tournes published three years
after the appearance of *Délie.* The poem of 1547, one of a kind, builds
on the areas in *Délie* where potamography is yoked to a gamut of psychic and erotic expression. A dialogue between two solitary souls (one
from Lyons and the other of pastoral environs), *Saulsaye* tells of the
naming of itself and, obliquely, of the city of its origin. The two woodcuts that Bernard Salomon crafted—one explicitly for the poem, the
other perhaps and perhaps not, at least insofar as it appears at a similar
juncture in Marguerite de Navarre's *Les Marguerites de la Marguerite
des princesses* (1547)—are so decisive to the force and form of the poem
that they cause it to be of a composite and almost coauthored design. In
the proximity of the juncture of the rivers of *Délie,* two shepherds speak
intimately with each other. Philerme, who could be a partner, companion, lover, or friend of Antire, is despondent over the loss of one of his
loves. He confesses that he has taken refuge in a solitary place where he
is free from the presence of two women, first Doris and then Belline,
for whom his love is unrequited. He is "disassembled," in disarray, in
amorous and psychic torment:

> Ainsi voyant ma totale ruyne,
> Deliberay de tout de m'absenter
> De sa presence, aussi de m'exempter
> De peine, ennui, cure, & solicitude:
> Et vins icy en ceste solitude
> Pour resjouyr quelque peu mes esprits,

Qui tant estoyent de mortel dueil surpris,
Ou me sentant loing de nostre paroisse,
De peu à peu j'amoindris mon angoisse.
Car le matin je vois là, ou la Saone
Vient à se joindre à son espoux le Rhosne,
Et le contraint à roidement courir
Jusqu'à la Mer, ou tous deux vont mourir.[17]

[Thus seeing my total ruin,
I immediately thought of absenting myself
From her presence, and of being exempt
Of pain, bother, disquiet, and solicitude:
And came here, in this solitude
To enliven somewhat my spirits
That were taken by mortal grief
Where, feeling my way far from our parish,
Slowly I reduced my anguish.
For in the morning I see there, where the Saône
Comes to be joined with its husband the Rhône,
And constrains it to flow steadily
To the Sea, where together the two will die.]

Antire plots the elegiac geography of the poem. He has left populous places to come here, *icy,* where in the light of morning he sees there, *là,* the two rivers in amorous confluence and congress. In the space of two lines the vision of a couple's life is imagined born and flowing to their demise. His words place him at the foot of the city of Lyons, at least in view of Salomon's woodcut image (Figure 32) on the opposite folio, set between the title (*Saulsaye* in 18-point roman majuscule) and its designation (*Eglogue* in 12-point roman majuscule). Set above the first five lines of the poem (inaugurated by a historiated *N* of "Non"), these three units are placed above the names of the two interlocutors who are given to be the dramatis personae. The woodcut (2¾" x 2¹⁄₁₆"), perhaps one of Salomon's most balanced images, depicts the place that Antire will soon describe.[18] The shepherds, in modern dress, are shown looking, perhaps admiringly, at the city of Lyons below the slope of Mount Fourvière in the distance. Turned away from the viewer, they sit on a flat spot to the right of a placid body of water (indicated by a tiny skiff floating in the distance). The shepherd to the left, arched on his knees, points skyward toward a church of two towers on the lower of two large slopes. The shepherd to the right, his left arm squeezing a bagpipe, seems to follow his companion's gesture with his eyes. The sheep graze peaceably in groups in the lower left and right corners of the scene while

SAVLSAYE·

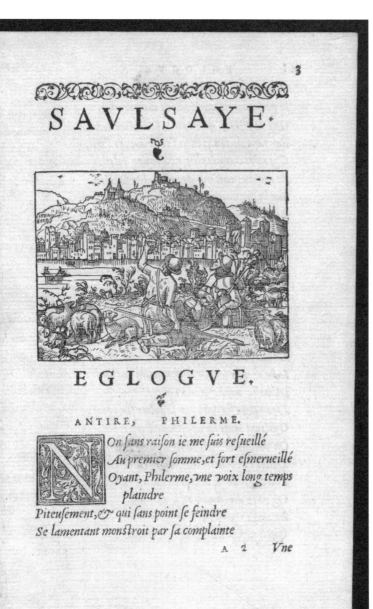

EGLOGVE,

ANTIRE, PHILERME.

On fans raifon ie me fuis refueillé
Au premier fomme,et fort efmerueillé
Oyant, Philerme, vne voix long temps
plaindre
Piteufement, & qui fans point fe feindre
Se lamentant monftroit par fa complainte

A 2   *Vne*

Figure 32. [Maurice Scève], *Saulsaye* (1547). Incipit and city view (woodcut) by Bernard Salomon, fol. 3r. [Typ 515.47.772] Houghton Library, Harvard University.

a faithful dog, resting to the left of the shepherd who holds his hand to the sky, turns its head in contrapposto to heed his master's gesture. In the corners to the upper left and right the sky is dotted with x-like incisions that indicate a calming presence of birds.[19] Particularly arresting is an implied vertical line that almost divides the woodcut into two equal sides: an axis can be drawn through the cut if the base of the fleuron below the central o of "Eclogue" serves as one plot point and the larger fleuron above (and also bisecting the title above it) another. The church-tower touches the upper edge of the cut and points to the lower curl of the fleuron above the frame. The implicit line leads the eye from the outside of the woodcut into its ostensible window, from the space the shepherds occupy in the foreground to the city view in the distance, and from its summit to the title and even the floral band above. The reader's eye looks *into* and is drawn *out of* an equally fictive and a real space. The pendant shape of the two fleurons causes the eye to descend into the center of the "Eclogue" to which the shepherd beckons as he looks upward. The eye is drawn both to the city and its position in respect to the title. The long staff he holds, which cuts diagonally across the image, can be taken to be more than a simple crosier. Because his left hand is raised to indicate something—but what?—on the horizon, and because, in the flattened perspective that encourages a haptic reading of the image, the hand seems to be touching the buildings in the distance. It can be read as raised to measure the distance between itself and the city beyond. The staff would thus be a variant on a surveyor's theodolite, and his hand raised as if it were holding a pair of dividers. In this way the shepherd would resemble a topographer charting the landscape in view of the image he will draw—but in which, simultaneously, he figures.[20]

In this way the woodcut that presents the dialogue inaugurating *Saulsaye* includes reference to its production of space. Commentators have shown that the companions are seated "nearly at the site of today's Faculty of Letters" at the University of Lyons, from where the view of the Cathedral of Saint-Jean and the hillside of Mount Fourvière are shown in geographical exactitude.[21] If so, by a short leap of the imagination the reader is enabled to see in the image not only the personages of Antire and Philerme, distracted from their melancholy dialogue, who measure and admire the city view of which they are a part, but also the poet and the artist in an "exurban" setting, a place outside an expanding city and not yet quite in the country. They cannot be said to be urban planners but, rather, visual shifters who provide points of reference between city and country and classical and modern landscapes.

The second woodcut in the work (Figure 33) underscores their function. Inspired in part by the almost futuristic effects seen in Francesco Colonna's *Songe de Poliphile* (1546, fol. 61v), here human bodies metamorphose into trees; the image is of a timeless landscape—a high and rocky promontory in the left and central background, lush vegetation below and to the right, the shore of a serene river hatched with pencil reeds at the bottom and left, to which four sets of figures are juxtaposed: in the center, a bevy of five nymphs holding each other's hands; to their left a pack of four satyrs who approach them, the leader playing a flute; in the foreground below, four satyrs (their furry legs and cloven feet identifying them), their arms outstretched, racing to the right, where, in the lower central area by the water's edge seven nymphs are in different stages of transformation into willow trees. The nymph to the left is a trunk and a leafy head that arches over the water, the next two leaning toward her, still of human shape; a third, in the clutches of a fourth nymph, who raises her arms in desperation; a fourth to the right who resembles the shape and pose of the first; two others, in the immediate background, whose heads display the drooping branches of the trees that will soon become the grove announced in the title. The artist takes such care to distinguish mobility from immobility that three nymphs in advanced stages of metamorphosis cast shadows over the water, while the one whom a satyr tries to violate leaves no trace of her profile on the surface.

The place *where* the woodcut illustrates the text is crucial to the composition of the poem. Philerme has conveyed his feeling of melancholy to Antire, who warns his companion that idle life in the country, far from good company, exercise, and sociality, may be noxious. Despite the beauty of the willows he sees, he asks, "Mais & comment pourrais tu vivre allaigre / En ce lieu cy, qui est de plaisir maigre? (fol. 12v, lines 224–25). ["But how could you live fresh and alert / in this place which is of meager pleasure?"] Good interlocutor that he is, Antire—whose name suggests both a lair *(antre)* and the act of withdrawing from it *(en tire . . .)*—will offer a narrative remedy. He will tell Philerme how the willows came to be where they are. But first the shepherds must "choose an appropriate place to stretch ourselves in leisure, away from the beating sun and, while our ewes ruminate in the shade, Hardy, your strong dog, will watch out for hungry wolves" (fol. 12v, lines 230–37). They repair to

Et la Palombe, aupres qui se contriste:
L'aure, & le vent y sifflent, & frisonnent:
Et tout autour les vndes y bouillonnent:
Ou ma voisine Echo t'escoutera,
Qui apres toy tes beaux vers chantera.

A N T.   *Vn iour parmy les Genetz verts flouris.*
Maints Dieux ensemble, & en ce lieu nourris,
Ioints auec eux Satyres demychieures,
Faunes aussi trop plus legers, que Lieures,
Et les Syluans hideusement cornuz,
Et la plus part pour le chault demy nuds,
Hors de tout hasle estoyent tous à l'vmbrage
Là, ou la Saone engresse son riuage,
Plaisant repos des prochaines forestz,

B 3   *En*

Figure 33. [Maurice Scève], *Saulsaye* (1547). Ravishment of nymphs, woodcut on fol. 13r. [Typ 515.47.772] Houghton Library, Harvard University.

the shade of a leafy elm where the widowed turtledove voices its saddened song, and where Echo, listening to Philerme, will redound his verse. Here the woodcut falls in place: at the site of the words to follow, it recounts three successive events—the nymphs coming forth into a clearing; the satyrs who happen upon and seduce them; and their near-ravishment and transformation, thanks to Arar, the god of the river (not seen in the image), who surges out of the waters to transform them into a willow grove—that are seen in simultaneity. The woodcut locates the beginning of the story in the poem by being set immediately above Antire's story within the poem. The mountain and the background could be Mount Fourvière, *avant la lettre*, in a classical landscape. The view would be located by its relation to the site the shepherds occupy at the beginning, especially in that the title anticipates the advent of what is seen in the second image. The timelessness of the decor in the second cut is broken—*desassemblé*—by the temporal notation in Antire's fairy tale begun below the lower edge of the scene Salomon has crafted: "*Un jour* parmy les Genestz verts flouris . . ." (fol. 13r, line 246; emphasis added). ["One day amidst the verdant flowering bloom . . ."] On *this* day an event brought a name to the place, *there,* "Là, ou la Saone engresse son rivage" (fol. 13r, line 253) ["where the Saône fertilizes its shore"]. The tale that follows, of 168 lines, is engraved on the landscape, and such that the presence of the surveyor or topographer is not entirely forgotten. The satyrs play their songs by pressing their fingers on holes "distantly compassed *[compassez]* of instruments sealed with tender wax" (fol. 14v, line 261) in order to lure the nymphs into view. When they dance to the satyrs' tunes they flail their arms in delight and throw their bodies into the air in a dance, "in moving themselves by alternate steps *[pas]* / without separating their measured compass *[leur mesurée compas]*" (fol. 16v, lines 330–31). The musical instruments pertain as much to sound as to the space of which they are a measure. In the description of the nymphs' seduction and plight the poem can be seen addressing the image that seems to authenticate the myth being told, and vice versa. No anteriority of either voice or image is evident. The nymphs fill the forests—seen in the cut—with their strident cries. They cannot escape the satyrs because the mountain behind them is steep and without a path or trail of egress. The only narrative element not included is Arar's eruption from the river, "his beard and hair soaked, / twisted in grass and reeds" (fol. 18v–19r, lines 384–85). The flat surface of the water becomes the mirror of the metamorphoses and holds the scene in a state of melancholic suspension.

The Country and the City

The moral that Antire draws from the story, if not the quasi-emblematic subscription to the image and its text, is that the willow grove can be a site of suicidal despair. The *saulsaye* is tied to the *lieu solitaire* and to deathly temptation resulting from solitude. The lazy lilt with which Antire tells his tale reflects the very idleness he decries. In all events Philerme has to "put the bridle to his teeth" (fol. 21r, line 454) to avoid the peril of despair. It is, literally, *désespoir,* what might in the memory of *Délie* bear the hieroglyphic form of a world or a sphere of emotion: come to your senses, look at where you are, and locate yourself with respect to the greater frame of things. "Vien donc mettre ordre en ton petit mesnage, / Et finiras en liesse ton aage" (fol. 22v, lines 468–69). ["Come now, when you finally put order in your household / In happiness you'll surely grow old."] For the sake of the dialogue, and perhaps for its dialogic relation with its "source" in Sannazaro, along with that of the Saône, Philerme can only alter his friend's counsel. The little *mesnage* or household that he secretly shares with Antire is not enough to sustain him. In view of the memory of the lost mistress Hope is an ardent purgatory. Antire adds that many means of diverting and deflecting passion are available in the city, and there also is found occasion for modest gain of capital. Philerme retorts that the amassing of money and fame in a public sphere hardly amounts to virtue or prudence. Antire: It keeps idleness at bay and can lead to good deeds. Philerme: We live with reason, and reasonably too, in solitude, especially right here, *en ce lieu.* Antire: The green willows, a sign of horror, bear fruit no more than idle souls who live and die in despair and folly. Philerme: He who dies here, with simplicity as his companion, is free where he would not be in the city; nature responds to sadness as if to one of kin, the bushes and foliage being "secret friends" (fol. 26v, line 574). Please note, replies Antire, that cold and snowy seasons hardly make for such bliss. Yes, surely, says Philerme, pastimes in cities are many, but so also are worries. Time and space are compressed—"le moys entier ne dure point un jour" (fol. 27r, line 594) [a whole month does not last a day]—while here pressures are fewer, choices greater, and pleasure more visible. Oh, please, quips Antire, a person who acquires wealth exploits generosity to gain favor. In a last and long exhortation that turns into an inner monologue until the final words of the poem, where interlocution gives way to meditation, Philerme muses, as if having Anteros counter the force of Eros, "Mais bien celuy est tres-semblable aux Dieux, / Qui aux honneurs clost, & bende les yeux" (fol. 28v, lines 626–27). ["Happy he who resembles the Gods / who refuses honors and covers his eyes."] That soul has not solicited *(solicité)* vainglory, living peaceably "in his innocent life," far from cities and their tumultuous and voluptuous ways. Intoning recall of Philippe de Vitry's famous "Franc Gonthier," the country poem for which François Villon penned an urban counterpart, Philerme takes solace in the idea of *not* having to curry favor in high places. He prefers his "rural" knowledge *(science)* to studied wisdom. He is far from hate, deception, and ambition, devoid of fear and the false expectations of hope. Delivered from desire, he lives happily in "sa logette honneste" (fol. 29r, line 668) [his honest little dwelling], in what might be his fabled farmhouse under a thatched roof. Even rustic objects undergo a happy metamorphosis:

> Les cailloux ronds luy donnent feu tisé,
> Les fleuves vin avec la main puisé,
> La terre pain, Arbres fruit, Chievres laict.
> De quelque tronche, ou lieu peult estre laid
> Luy sont le miel tresnet, & copieux,
> A savourer doux, & delicieux. (fol. 30r, lines 688–93)

> [Round pebbles give spark for his fire,
> Rivers wine drawn from his hand,
> Earth bread, Trees fruit, Goats milk.
> From some trunk, or perhaps ugly place
> For him flows clear and copious honey
> To savor its sweetness and delight.]

His eulogy of nature almost turns into an anthropomorphic landscape: streams are noisy, argentine, and fluid; rocks are mossy and caverns humid; woods flourishing with "poignant" wild roses and hawthorns embalming the paths and trails. Rocks respond to his songs, the hum (or white noise) of honeybees puts him to sleep. When he wishes he reposes in the nude, in pleasure unknown to the clothed life of the city.[22] And yet the unknown is felt in the shadows:

> Heureux Pasteur, si se peult dire heureux
> Homme, qui vive en ce val tenebreux. (fol. 31r, lines 714–15)

> [The happy Pastor, if he can be said happy,
> the man who might live in this shady valley.]

The praise of happiness appears scored with doubt much as the valley in penumbra. At that moment in the dialogue Philerme decides to stay where he is, "en ce lieu sans plus les villes suyvre" (fol. 31r, line 719) ["in this place, here, no longer seeking the cities"]. He turns toward Antire, addressing him directly, inviting him to spend the night together before they disband. While he regathers his sheep he asks Antire to pick up his sack and bottle before they go off to sleep.

The last four lines of the poem refer directly to the composition of the first woodcut. The final and pivotal word of the poem is the toponym of the mountain toward which the shepherds had been seen gazing:

> Voy tout autour le Daulphiné à l'umbre
> Pour le Soleil, qui delà la riviere
> S'en va coucher oultre le mont Forvière. (fol. 32v, lines 728–30)

> [Behold the Dauphiné all about in the shadow,
> For the Sun that past the river
> Goes off to set beyond Mount Fourvière.]

The elegiac and often conventional matter of the city-country debate is ultimately situated in a place of both ending and origin. In their field of view they gaze upon the hillsides of the Dauphiné in the growing shadows "going to sleep" with the sun as it sets over the mountain. The shepherds, nature itself, and the poem fall into slumber, at the very site where the world had awoken. The drift of Philerme's words suggests that the story of the nymphs, together with the debate that follows, serves as a cure in the strongest psychic sense of the word. The conversation brought him to a conversion, to a point where he has "worked through" his grief and—at least through the filter of the printed words of the poem—now identifies his place in the world in respect to locating Mount Fourvière in the distance (in the narrative) and in the immediate present (in the sight of the woodcut and the shape of the book). When FIN falls below the place-names, the mountain and its environs become a vanishing point and a convergence of the text and the inaugural image.

It is difficult not to see in Salomon's woodcut the location of a collaborative endeavor that includes the labors of the artist, topographer, poet, and engraver. *Saulsaye* situates the ferment of Scève's work in a specific geography that produces new spaces: not utopian, such as those of the early Rabelais's Abbaye de Thélème, but those that welcome meditation on the nature of living and desiring in relation to objects, people, and places. If Scève belongs to the fabled *École lyonnaise* that brought poetry into new directions in midcentury France, at a moment when the nation is on the cusp of religious turmoil, its sense of itself—its origins, its heritage, its inventions in the spatial arts—makes it a point of reference for other programs that tie topic and place to the space of verse. The poets of the Pléiad are of course the first and most famous poets to draw upon *Délie* and *Saulsaye*. It remains to see how one of them, Ronsard, fashions poetic topographies of different signatures and places.

5
Ronsard in Conflict: A Writer out of Place

One of the tasks of the two preceding chapters was to determine some spatial and topical modes of creation in books of emblems and poems in which woodcut images are found. It was observed that emblems draw on cartographic form and develop spatial modes of their own, and that in the hands of Scève and his collaborators poetry is altered to generate intense speculation about the desiring self, about where it moves in the ambient world, and the nature of its relations with cartographic idioms. In the 1530s and 1540s, innovation and experiment with the printed and illustrated book appear to give reason to the spatial force of poetry. Lyrical verse, we have seen, is hieroglyphic, and hieroglyphs as they had been affiliated with ancient Egypt were said to belong to a latent or manifest language of geography. The creative drive of poetry of the moment is so often likened to engraved images that verbal material, much as it is in contemporary maps, acquires specific spatial and iconic densities.

The aim of this chapter is to follow this line of inquiry through Pierre de Ronsard, who, more and better than any other poet in sixteenth-century France, takes great care to establish a world map and a territory of his own signature. In his early verse explosive invention is tied to topographical imagery, on the one hand while, on the other, the later matter appeals to the recent memory of the emblem and of woodcut images. As the poet confirms his signature and forges ahead, his name

is cast in greater doubt. Now and again the space it had arrogated for itself is called into question. When national fortunes change so do those
of the poetic topographer who anxiously, perhaps because of a heightened awareness of things unknown, situates his ego in a broader space
of time and being. Where he does he inquires of his own alterity with
respect to the world in which he resides.

Like Scève, Ronsard is the subject of a critical industry. Rightfully recognized as the greatest French poet of the sixteenth century (Ronsard,
hélas, as adepts of Scève, Du Bellay, Louise Labé, or Agrippa d'Aubigné
might say, miming André Gide's estimation of Victor Hugo), his verse is
varied and protean. Most of the poems witness endless and often febrile
modification and alteration for the duration of Ronsard's writing life.[1]
Rethinking his verse as he goes, from the *Odes* up to the posthumous
edition of the *Oeuvres complètes,* he seems to conceive his poems as a site
of metamorphosis. For the historian its geographical latency is not sur-
prising. In his lifetime Ronsard had received a commission, with Jean
Daurat, Jean-Antoine de Baïf, and Nicolas de Nicolay, for belonging to
an elite group of "'docttes geographes, historiographes et traducteurs'"
(Pelletier 2009, 35) [learned geographers, historiographers, and transla-
tors]. His library, too, contained works of classical geography, like that
of *Stephanus de urbibus* (Figure 34), on which he penned his signature,
that in most likelihood served as guides for his construction of verbal
landscapes of both contemporary and ancient allure. Through the art
of geography the poems become spaces of endlessly local and total invention, sites where *topos* and *cosmos* are intertwined. For the sake of
plotting, the discussion that follows will begin with locale and creation
in the *Amours* of 1552 and 1553, particularly where the poet marks or
engraves on the page, in the sense of an inscription on wood or copper, any number of locational figures. Poetic turns and formulas, not
unlike those encountered in Scève, are affiliated with a locale that can
at once be the area to which tactical reference is made (as in his own
signature, "Pierre de Ronsard Vandomoys"), the book or poem itself,
and the situation in which the poet finds himself in the welter of current events. The locale changes as it resituates the relation of the poet
to a world whose borders, as was seen in Rabelais (chapter 1), extend
outward. When they do, the possessive and egocentric nature of the
poetic persona becomes, even in the most political verse that defends
the established order, paradoxically ethnographic. Such is where topography includes alterity in its field of inquiry. Within the imaginary

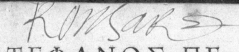

ΣΤΕΦΑΝΟΣ ΠΕ-
ΡΙ ΠΟΛΕΩΝ·

STEPHANVS
DE VRBIBVS.

GVILIELMI XYLANDRI AVGVSTA-
ni labore à permultis fœdisq; mendis repurgatus, duo-
busq; Inuentarijs (uno Autorum,quorum è fcriptis
teftimonia petuntur:altero,Rerum & uerborum
memorabilium) auctus.

Cætera ex eius ad Lectorem Præfatione
intelliges.

Idem opus Latiné factum,infertum eiufdem X Y L A N D R I
Onomaftico Geographico, paulo pôft in
lucem dabitur.

Cum Cæf. Maieft. gratia & priuilegio
ad annos fex·

BASILEÆ, EX OFFICINA OPO-
RINIANA. 1568.

Figure 34. Title page of Ronsard's copy of *Stephanos per poleon* [=] *Stephanus de
urbibus* (Basel, 1568), a book of names and places in classical geography. [f FC5 R6697
Zz568s] Houghton Library, Harvard University.

world where the self is free to wander errantly (*libre follastre où le pied le conduit,* free to wander where its foot leads it), whether into the classical past or into contemporary events, it considers where it is in the environing world. To see how these imaginary worlds shift and how their topographies are in mutation, three arenas will be taken up: the *Amours* of 1552, the *Continuation* of 1555, and the mixed and difficult verse that moves from self-doubt and consternation in 1559 to the polemical *Discours* from 1562 to 1565.

The *Amours* of 1552, which included the *Cinquiesme livre des odes,* under the printed name of "P. de Ronsard, Vandomois," includes 188 poems, the greater part sonnets, exploiting the recent success of the form since the experiments of Mellin de Saint-Gelais and Marot, on the one hand and, on the other, Vasquin Philieul's *Laure d'Avignon,* a strong translation of much of Petrarch's *Canzoniere* (1547) and Joachim Du Bellay's *L'olive* (1549/1554). The dazzle and panache of the *Amours* prove that Ronsard was not to be outdone by anyone, whether his contemporaries or even Petrarch, after whom he fashions himself in word and image. The woodcut frontispiece to the first edition (Figure 35) is patterned after the cordiform frame in which Petrarch and Laura are shown in various editions of the *Opere volgari* (Alduy 2007, 309–53). In an oval surround and in profile, looking right, he beholds Cassandre, a Fontainebleau variant of Laura on the opposite folio (Figure 36). Dressed in a cloak attached to an oval pendant of a necklace, her breasts bare, in profile she looks toward the poet. Beneath an ample coiffure tightened by a jeweled clasp, her elongated neck draws attention to two ringlets of hair that waft in the white space behind her. A smooth complexion draws attention to the Roman beauty of her profile. Ronsard, also of aquiline aspect, wears a closely cropped beard and is adorned with the branch identifying him as a poet laureate. A muscular neck and shoulders emerge from the toga draped over a robust upper torso. As in the emblematic tradition, the two figures are seen in a surround that displays curled strapwork at the top and bottom of each oval, distantly recalling the motif of Corrozet's enigmatic snail and the spirals in the emblems of *Délie.*

A Graven Style

The details of the frontispiece recur in the text. The parts of Cassandre's body are the topic of several sonnet-blazons, including her hair, face, neck, eyes, and lips. She looks almost with admiration at the poet, whose gaze is so assertive that it seems to survey the landscape of the woman's face and torso, of a beauty soon said to astonish and bewilder him. The double portrait of the frontispiece, the only illustration of the text, nonetheless underscores the *grave* style that Ronsard will use with oblique reference to the art of inscription and epigraphy. Now and again in the sonnets he stages the scene where he cuts a signature into the bark of a tree in his native Touraine:

> *Quand ces beaulz yeulx iugeront que ie meure,*
>   *Auant mes iours me fouldroyant la bas,*
>   *Et que la Parque aura porté mes pas*
>   *A l'aultre flanc de la rive meilleure:*
> *Antres & prez, & vous forestz, a l'heure,*
>   *Ie vous supply, ne me desdaignez pas,*
>   *Ains donnez moy, soubz l'ombre de voz bras,*
>   *Quelque repos de paisible demeure.*
> *Puisse avenir qu'un poëte amoureux,*
>   *Ayant horreur de mon sort malheureux,*
>   *Dans un cyprez notte cest epigramme:*
> CY DESSOUBZ GIST VN AMANT VANDOMOYS
> QVE LA DOULEUR TVA DEDANS CE BOYS:
> POVR AYMER TROP LES BEAVX YEULLX DE SA DAME.[2]

> [When these beautiful eyes will judge that I die
>   Before my days thrusting me down,
>   And when Fate will have led my feet
>   To the other side of the better shore:
> Lairs and fields, and you forests, now,
>   I beg you, don't disdain me,
>   But award me, under the shadow of your arms
>   Some repose in a peaceful place,
> And there may that an amorous poet,
>   Horrified by my sad destiny,
>   Note this epigram on a cypress:
> HERE LIES A LOVER FROM THE VENDÔME
> WHOM PAIN KILLED IN THESE WOODS:
> FOR LOVING TOO MUCH HIS LADY'S WONDROUS EYES.]

In his feigned grief he locates himself and his love on a cypress— appropriately chosen because of the density of the wood and because it is *ci-près,* right here—in what he calls an *epigramme,* the substantive that Alciati had used to define an emblem. Wanting to choose an ideal site for his sepulture, "A l'aultre flanc de la rive meilleure" (63, line 4, 41) [On the other side of the better shore], the Vendômois poet's allusion to the Styx suggests that the river could be the Loir. It is where, in a conventional pose, he begs, "Antres & prez, & vous forestz, à l'heure"

Figure 35. Portrait of Ronsard in the 1552 edition of *Les amours*. [FC5.R6697.552a] Houghton Library, Harvard University.

Ὡς ἀπὸ ρωνσάρδ'ϑ
εἰς τὴν Κάσανδραν.
Φοιβάδα τὴν Κάσανδραν, ἔρως τὸν ἕταιρον ἐκείνης,
Φοιβομανῆ τεῦξεν φοῖβός ἐρωμανέων.
Ἡ δ'ἄλλη Κάσανδρ'ᾗ ᾗ κελπίδος, οὐκέτι φοιβάς,
Νῦν ἐμ'ἐρωμανία ρέξ ἰδὲ φοιβομανῆ.
Ἰα. Αντω. Βαίφιϑ.

Figure 36. Portrait of Cassandre, facing Ronsard, in the 1552 edition of *Les amours*. [FC5.R6697.552a] Houghton Library, Harvard University.

(line 5) [Lairs and fields, and you, forests, quickly] to award him a resting site. In its printed form the formula signals a topography. Less important than the commonplace of the election of a rustic tomb for a pining lover (of twenty-seven years when the *Amours* were first published) is the allusion to the production of lexical effects that multiply and reiterate a sense of place. The cypress in which the words are incised can be taken to be the paper page of the book before one's eyes. The *epigramme* calls attention to the visual and lexical disposition of the poem. Something of a poetic map, its writing is a function of its deixis (stating where it is situated and how) and its own composition.

The lowercase majuscules of the final tercet form a base on which the two quatrains and first tercet repose and are the end point of a narrative itinerary. More than that, they attest to different graphic speeds in the same piece. The elegant italic typography, which moves forward with an effect of alacrity (the 1552 edition deploys two z's with longer and shorter serifs) is shown in opposition to the classical style that would belong to the epigraphy found on tombstones. The epigram prints V where the corresponding italic above is u. The writing of the first two quatrains and the initial tercet gives way to a writing in which the Latin V is shown in strong contrast to its lowercase counterparts above. The letter has a visual valence that brings forward the masterful typographic design of the edition of 1552 on which the recto and verso sides of each folio print two sonnets in their integrity. The architecture of the book is enhanced, and so also the integrity of each sonnet-piece: no numerals, either roman or arabic, are set over the poems as they are in modern editions. As in *Délie,* the spatial syntax dictates that they be seen in units of two or four per page or per folio. Further, the incipit *Quand ces beaulz yeulx* establishes not only a visual and cultural difference between its shape in the first line and its reiteration in the last but also a virtual link in a concatenation reaching back to the *Voeu* of the beginning, and forward, to the optical aspect of the sonnet just below:

> *Qui vouldra voyr dedans une ieunesse,*
> *La beaulté iointe auec la chasteté,*
> *L'humble doulceur, la graue magesté,*
> *Toutes vertus, & toute gentillesse:*
> *Qui vouldra voyr les yeulx d'une deesse,*
> *Et de noz ans la seule nouueauté,*
> *De ceste Dame oeillade la beaulté,*
> *Que le vulgaire appelle ma maistresse.*

Il apprendra comme amour rid & mord,
  Comme il guarit, comme il donne la mort,
  Puis il dira voyant chose si belle:
Heureux vrayment, heureux qui peult auoyr
Heureusement cest heur que de la voyr,
  Et plus heureux qui meurt pour l'amour d'elle. (ed. 1552, fol. 31r;
    LIV, 55; PLI, 56)

[Whoever wishes to see in a youth,
  Beauty joined with chastity,
  Humble sweetness, grave majesty,
  Every virtue and every gentility:
Whoever wishes to see the eyes of a goddess,
  And of our years the sole novelty
  May he gaze upon the beauty of this Lady,
  Whom the common call my mistress.
He will learn how love laughs and bites,
  How it cures, how it kills,
  And then he will say in seeing such a handsome thing:
Happy, truly happy whoever can have
  Happily this bliss to behold her,
  And happier he may die for her love.]

Reiteration of the beginning of Petrarch's *Canzoniere* is clear, and so is the sense of a stunning body transformed into printed shape on a paper surface. *"L'humble doulceur, la graue magesté"*: the poem has spatial valence in making both aural and visual reference to the dedicatory sonnet that shifts between the frontispiece and the poems below. In that poem the title, V OE V, is set above a sonnet that both inaugurates and continues the "grave" style associated with inscription, impression, epigraphy, and the geography of beloved bodies:

V OE V.
Divin troupeau, qui sur les rives molles (fol. 3v; L4, 4; PLI 19)

not only draws the eyes to its engraved aspect but also bears the faint image of Cassandre in the oval on the other side of the folio. The dedicatory sonnet *reengraves* what is drawn to the eye in its figure of a monogram: a *voeu* and a vow, but a sign or even a hieroglyph pregnant with meaning, like an egg *(oeuf)*, one of the icons of creative perfection in the *Amours;* a collection to be seen, *veu,* from its pictured beginning; an assemblage compassed and engraved, by way of the origin of sight in the angular *V* that heralds the homeland, the *V . . . andomoys,* of the author. He invokes the nine Muses on the *rives molles* (line 1) [soft shores]

of the Euroteus, both a real and a mythic river whose course might
be the line that a burin cuts in drawing its course or that of the poet's
inspiration, which goes from Mount Parnassus (in the same line) to
the Fountain of Hippocrenes and the stars whose configuration at his
birth not only destined him to become a poet but also sealed his fate
in the finality of a woodcut image. He asks the Muses, *"Plus dur qu'en
fer, qu'en cuyvre ou qu'en metal / Dans vostre temple engravez ces paroles"*
(lines 7–8). [Harder than in iron, than in copper, or than in metal / In
your temple engrave these words.] In the tercet (in small caps) *Ronsard*,
who asks the "divine troupe" of Muses to accord him a place—the
name figuring at the incipit—and his book at the foot of his *Idole*—the
word set at the cornerstone or *pierre angulaire* [keystone] of the poem.
The words refer to the frontispiece figuring Ronsard and his idol and to
the "engraved" style that situates the poetic topography.

The poet's melancholy is everywhere tempered where attention is
drawn to the incised or printed character of the verse. Because each
sonnet is a partial unit of an open and ever-expanding whole, the collection can be loosely compared to Petrarch's *Canzoniere* in which the
sonnets figure indirectly as part of a cartographic design, as an emotive
island in an autobiographical *isolario* or book of islands. The poems
can be "read" in terms of their projective design, both in the sense of
amorous identification and of plotted form. Many sonnets lead to and
away from a center that seems to be the origin, umbilicus, target, bull's
eye, vanishing point, or, in topographic terms, the axis of central lines
of latitude and longitude about which the sonnet is constructed.[3] They
respond in graphic terms to the Petrarchan invocation, *Qui voudra voir,*
the incipit of the first sonnet and the fifty-fourth, that which follows
both the invocational sonnet and the sixty-third below.

Ciel, air, & ventz, plains, & montz descouvers

The traits of Ronsard's style can be appreciated as recurring graphs or
marks in a pattern of difference, repetition, and variation of formulas that
convey Petrarchan, Ficinian, and other themes. In begging the forests and
fields to accord him a resting place, the poet hardly seeks repose. The formulas in *Je vous supply* and in *Qui voudra voir* redound in the *Amours* and
thus make possible a transverse reading of the sonnets, and notably the
sixty-seventh (fol. 43r). One of the more conventional in the collection, it
also one of the most exhilarating:

Ciel, air, & ventz, plains, & montz descouvers,
  Tertres fourchuz, & foretz verdoyantes,
  Rivages tortz, & sources ondoyantes,
  Taillis razez, & vous bocages vers,
Antres moussus à demyfront ouvers,
  Prez, boutons, fleurs, & herbes rousoyantes,
  Coustaux vineus, & plages blondoyantes,
  Gastine, Loyr, & vous mes tristes vers:
Puis qu'au partir, rongé de soing & d'ire,
  A ce bel oeil, l'Adieu ie n'ai sceu dire,
  Qui pres & loin me detient en émoi:
Ie vous supply, Ciel, air, ventz, montz, & plaines,
  Tailliz, forestz, rivages & fontaines,
  Antres, prés, fleurs, dites le luy pour moy.[4] (fol. 33r; L4, 59; PLI, 57–58)

[Sky, air, & winds, plains, & uncovered mountains,
  Cloven hillocks, & greening forests,
  Twisting shores, & undulating springs,
  Cropped thickets, & you green groves,
Mossy lairs opened halfway,
  Fields, buds, flowers, & dewy grasses,
  Veined hillsides, & fields shimmering in sun,
  Gastine, Loir, & you my sad verse:
Since your departure, fretting with fear and ire,
  To this handsome eye, I knew not how to bid adieu,
  That near & far keeps me in chagrin:
I beg of you, sky, air, winds, mountains, & plains,
  Bushes, forests, shores & fountains,
  Lairs, fields, flowers, say it to her for me.]

Je vous supply triggers (indeed, supplies!) the hastened return of all the elements enumerated in the two quatrains above. A style of form with many variants in Ronsard's wit, it moves with the velocity of an eye scanning a landscape. The sonnet constructs the area that the eye touches at the speed of its reading, a sonnet whose italic form bends forward, as if blown by the wind of the inspiration that utters it. Said to be inspired by a sonnet by the well-named Astemio Bevilacqua of 1547 (Weber and Weber 1963/1998, 534), in its own context it belongs to the cultivated countryside of Ronsard's Vendômois and the fields of figures depicting similar places in other sonnets in the collection.[5]

Here, however, the relation of the cosmos, as *Tout* or an undecipherable whole and a mass of topographic details—passing images, hieroglyphs, or emblematic forms—comes forward from the very beginning. Five attributes of the landscape are shown in the first line, from the cosmographic horizon, the sky with its implied "celestial vault," and the agitation of atomic particles in the air and winds witnessed before a deceptively simple opposition of flat and hilly horizons ends the decasyllable. When seen and spoken simultaneously, *montz descouvers* can be understood as "uncovered mountains," but also as *mondes couverts,* covert, covered, hidden, secret worlds of great and even total extension within the details in which they are concealed. In the same line a balance between two scales is given in the setting of ampersands, each an instance of speedwriting to accelerate the verse and to bring forward recall of a *festina lente,* a visual oxymoron in which curved and straight lines are in coincidence. Set as if in the measure of the third and sixth beats of the line, the sign signals that the eye does well to survey the space of the poem. No single sonnet of the *Amours* is so riddled with the ampersand, and none so visually sets its quatrains so evidently into mirrored halves. In line 2 the ampersand falls after the fourth syllable and likewise in lines 3, 4, 6, 7, and 8.

The splitting or (à la Scève) "disassembling" effect seen when the eye follows the crosswise itinerary in vertical scansion of the ampersand gives way to a reintegration of the twelve descriptive fragments of the landscape in the final tercet, where they are reiterated in lines 13 and 14. Division is present in the narrative of departure, while a sense of unity comes when the poem has "worked through" the grief (if grief there was) in its own plotting and writing. Where the poem uses the ampersand to signal its lines of divide it equally alludes to its engraved aspect. Critical is the vanishing line that begins with the formula *Coustaux vineus,* partially derived from Du Bellay's descriptions of rivers in *L'olive,* which can be accurately pictured as "hillsides fertile in wine," that is, cultivated with rows of grapevines, which in topographic views imply slopes that have been "civilized" or managed in the hands of man. An icon in a lexicon developed to designate vineyards, one sign was a mark fashioned from straight and curved traits that resembled an ampersand (Dainville 1964/2002, 327, fig. 47; Delano-Smith 2007, 574, fig. 21.44).[6] The manicured hills would be a perfect complement to the golden plains shimmering in sunlight, *plages blondoyantes,* which resemble wheat fields in August as depicted by Niccolò dell'Abbate or Fontainebleau painters.[7] *Coustaux* are hillsides, but they are also cutting instruments, knives and burins that incise veins into the surfaces employed to represent these hillsides. The cutting edges that prune the vines yield their *vins.* Three sorts of images come forward: that of the cutting blade; slopes and hillsides; the products yielded, which are at once images, poetic lines, and graphic signs referring to the land with which the poetry is associated: Gastine (the woods close to his own home that would later be deforested), the Loir (the river that runs through it), and "you my sad lines," that is, the *vers* that lean toward those places affiliated with the poet's Vendômois signature.

Cutting edges are everywhere in the poem. *Tertres fourchus* refers to hills split by valleys, but also to the forkings or birfurcating effects in the landscape that can be seen in *tertres,* a substantive that splits in two to make two "threes" or *ters.* The rhymes swing on *vers—descouvers, verds, ouvers, vers*—with such resonance that the two quatrains are anticipated by the cosmographic formula, "Ciel, air, & ventz " at the outset, in which *air, & ventz* has scrolled in its form the sign of *vers.* Finally, the princely ego of the poet, his *moy,* is split by being both inside and outside itself, in *esmoy,* before, in the last word, it becomes the key to the vault whose groins are seen leading to the other side of the poem, in the inaugural *Ciel.*

The sonnet is about parts and wholes, and it is also about the force of hieroglyphic conjunctions of things, beings, feelings, and spaces. The magnificent *&* that hastens it to its end seems to touch on infinity in its force of addition. Never negating, never denying (as do Scève and Du Bellay in different ways), it accumulates images of partial and local space, one *and* the other after each other.[8] "Ciel, air, & ventz" mobilizes the more didactic (and frequently anthologized) poems that in the mantle of Lucretian atomism and the Epicurean *clinamen* embody the errancy of motion less than they write of it, as in the thirty-seventh sonnet, *"Les petis corps, culbutans de travers, / Parmi leur cheute en byaiz vagabondz / Hurter ensemble, ont composé le monde . . ."* (fol. 25v, lines 1–3]. [The little bodies, bouncing all about, / Among their fall in an errant bias, / Bumped together, composed the world . . .] The world made of loosely aggregated particles, much as the *Amours* its poems, is comparable to the author's *amoureux univers* (line 8) [amorous universe], indeed an affective geography, that is more or less tied—but the unknown secrets of the world are never divulged to say why or how—to *le grand Tout* (line 14) [the great Totality] that remains, also in the manner of the *Amours,* open and both liable and viable to expansion.

A Parting Shot

The engraved, plotted, and mapped character of the open whole is made clear in the 221st sonnet, which closes the edition of 1552. In homage to his patron Henri II, celebrating the king's triumphs of his campaign along the Rhine in the summer of 1552, Ronsard launches one of his usual comparisons by which the inferior term, through the process of its articulation, exceeds and wins over the better analogue to which it is compared. The poet sets the image of himself engaged in creation of idle fancy against the nobler endeavors of the heroic and glorious monarch battling for the cause of France at one of its contested borders:

> J'alloy roullant ces larmes de mes yeulx,
>   Or plein de doubte, ores plein d'esperance
>   Lorsque HENRY loing des bornes de France,
>   Vangeoyt l'honneur de ses premiers ayeulx.
> Lors qu'il trenchoyt d'un bras victorieux
>   Au bord du Rhin l'Espaignolle vaillance,
>   Ja se trassant de l'aigu de sa lance,
>   Un beau sentier pour s'en aller aux cieulx.
> Vous saint troupeau, qui dessus Pinde errez,
>   Et qui de grace ouvrez, & desserrez
>   Voz doctes eaux à ceulx qui les vont boyre:
> Si quelque foys vous m'avez abreuvé,
>   Soyt pour jamais ce souspir engravé,
>   Dans l'immortel du temple de Memoyre. (1552, fol. 96v; L4, 172; PLI, 156)

> [I went about rolling these tears from my eyes
>   Now full of doubt, now full of hope
>   When Henry, from the borders of France
>   Avenged the honor of his first ancestors.
> When he severed with a victorious arm
>   Spanish pride at the shores of the Rhine,
>   Then by tracing with the tip of his lance
>   A handsome path to go off to the heavens.
> You, saintly troop, who err over Olympus,
>   And generously open and release
>   Your learned waters to those who go to drink them:
> If sometime you have slaked my thirst
>   May this sign forever be engraved,
>   In the immortal hall of the temple of Memory.]

The scene where Henry traces the tip of his lance on the shores of his campaign is lifted from the *Georgics* and set upon the shores of the Rhine. At the end of the eighth line the meandering path it cuts into the ground, however it might lead to the sky, ends abruptly. An engraved figure has left its trace and made a map that runs from Caesar on the Euphrates before moving to more immediate history. It follows in the comparative gist of the two tercets that the poet seeks to follow a route to the airy lands where Muses wander errantly, and to find a sepulchre in immortal place in the temple—in the last word, set at the cornerstone of the collection, driving the point home—of Memory, where the mirrored *moy* (in Me-*moy*-re) will be eternized forever. He wants his book to be *engraved* therein, just as it had been, from the beginning, drawn from the embossed image attesting to a poetic and geographic genealogy.

The perverse turn that makes the king the poet's subject is met by the inscription of Henry's name in Roman majuscules. It is shown to be at the limits of the nation where, in the syntax of the fifth line, it is not clear to whom in the synecdoche the "victorious arm" that cuts the Spanish nation to size might belong—at least before the seventh line specifies that he is engraving the earth with it tip as might, it will soon be shown, *a poet with his burin.* For this reason the figure of the "I" who writes about he who went "rolling these tears from my eyes" (line 1) seems awry. Tears generally fall or roll *down* one's cheeks, not *across* them. Ronsard's *armes,* his artillery and ordnance, are his *larmes.* For all the immortality he gains at the inferred expense of the king whom he is paid to glorify, site and situation are immediate and contingent. They belong to a local place: as Jean-Paul Sartre might have put it, Ronsard is in an existential situation, *chez lui,* at his desk, plotting a way to immortality above and beyond that of his patron.

Ronsard Saved from Drowning

Two years later, at the very outset of the *Continuation des Amours,* following *Le bocage* (1554; in Weber and Weber 1963/1998, 143–53), whose title typifies the geographic imagination of the miniature collection, and his *Meslanges* of 1555 (ibid., 155–69), in addressing his friends and fellow members of the Pléiade, Ronsard acknowledges (or invents) the scene of a harsh public reception of the *Amours.* They were, he asserts, rejected on the grounds of his obscurity. He deploys "heroic verse" to elevate the virtue of his new mistress, no longer Cassandre, but now a rustic maiden named Marie de Bourgueil. He risks, he says, being accused of having too "low" a tenor of writing. What to do? Genius that he is, he can only become "this monstrous Proteus" in perpetual metamorphosis. In their feigned simplicity the twenty-seven sonnets in lines of twelve syllables, the eleven others in a measure of ten or eleven, and thirty more in "heroic verse," he seems to come home to roost in his native land. The collection is shown to be an exercise in *style,* of producing a multiple or protean signature in the new ways he twists and turns his words (Rigolot 2002a, 196). In the famous seventh sonnet he exploits,

following Du Bellay's praise of its virtues in the *Deffence,* the seductive
force of the anagram. *"Marie, qui voudroit vostre beau nom tourner, / Il
trouveroit Aimer: aimez-moi dons, Marie"* (Weber and Weber 1963/1998,
175, lines 1–2; L7, 123; PLI, 182) [Marie, whoever would wish to turn
your name, / would find To Love: so then love me, Marie], especially
since his love could never be written from a *better place* (line 4). He uses
the stratagem of turning the name in order to obtain her charms. He
also avows that *tourner* also means *trouver.* Turning about Marie, in
gaining a view all around her, he seeks to capture her in his panoramic
field of view. Yet he who turns and finds gets turned about and found:
the last sonnet in the collection praises his beloved for having turned
him and his style in new directions:

> *Marie, tout ainsi que vous m'avés tourné*
> *Mon sens, & ma raison, par vôtre voix subtile,*
> *Ainsi m'avés tourné mon grave premier stile . . .* (ibid., 213, lines 1–3)

> [Marie, just as you turned
> My rhyme and reason in every which way, by your subtle voice,
> Thus you turned my first grave style . . .]

The verbal pressure he exerts is controlled and local, not *abandonné*
(line 8) (in abandon) as it was when he was under the spell of Cassandre.

The cosmographic dimension of the *Continuation* seems attenuated,
and a sense of topography more pronounced. The nineteenth sonnet, in
which the dynamic of gaining an upper hand on the object he immortal-
izes through the writing, indeed the bent meander of his praise, is ori-
ented in the direction of geography. One day, while boating on the Loir,
the tributary of the Loire with which he continually affiliates himself (as
in sonnet 63 of the *Amours*), he and his craft were capsized. Forgetting the
rustic pleasure he would share with Marie, he invents a dialogue with the
river in order to chastise it for its fluvial effrontery:

> *Mais respons, meschant Loir, me rens-tu ce loier,*
> *Pour avoir tant chanté ta gloire & ta louange?*
> *As-tu osé, barbare, au milieu de ta fange*
> *Renversant mon bateau, sous tes eaus m'envoier?*
> *Si ma plume eut daigné seulement employer*
> *Six vers, à celebrer quelque autre fleuve estrange,*
> *Quiconque soit celui, fusse le Nil, ou Gange,*
> *Comme toi n'eust voulu dans ses eaus me noier:*
> *D'autant que je t'aimoi, je me fiois en toi,*
> *Mais tu m'as bien montré que l'eau n'a point de foi:*
> *N'es-tu pas bien meschant? Pour rendre plus famé*

Ton cours, à tout jamais du los qui de moi part,
Tu m'as voulu noier, afin d'estre nommé
En lieu du Loir, le fleuve où se noya Ronsard. (ibid., 183; L7, 136; PLI, 500)

[Now answer, naughty Loir: do you praise me
    For having sung so much of your glory and your esteem?
    Did you dare, barbarian that you are, in the midst of your muck,
    Capsizing my craft, send me beneath your waters?
Had my quill deigned merely to use
    Six lines to praise another foreign river,
    Whoever it be, whether the Nile or Ganges,
    Like you, it would never have wished to drown me:
For I loved you, I had faith in you,
    But you surely showed me that water's faith is feckless:
    Aren't you really naughty? To bring fame to yourself,
Despite all the praise I've rendered,
    You've sought to drown me so to be named,
    Instead of the Loir, the river where was drowned Ronsard.]

It has been noted that although Ovid scorns a rising river that impedes him from joining his beloved, this poem "was inspired by a real accident"; that in truth Ronsard had praised the Loir at length in two of his *Odes,* "Au fleuve du Loir" and "A la source du Loir." In various sonnets he associates the flowing waters with his pains of love (Weber and Weber 1963/1998, 620).

The water tables are turned. Ronsard bends an established toponym into his own name; he exploits the anagram to make the *Loir* serve the ends of tongue-in-cheek *loier,* in short, his art of crafting a paradoxical encomium. The image of a greater world and its rivers, two of which are enumerated in the seventh line, indicates the perverse measure of wit taken in the sixth line, when he is *not* using six lines to pay homage to their glory. The waters are shown to be as protean as the author and his verse—*l'eau n'a point de foi*—and thus mirror the poet who has changed his manner and style in the shift from Cassandre to Marie. The change of the lover's name is like that of the waters, "Instead of the Loir, the river where was drowned Ronsard." The surname is engraved at a crucial cardinal point in the geography of the poem.

Mixed Fortune

Often in the sonnets written and altered from the *Amours* to the *Second livre des meslanges* (1559) and to the first edition of the poet's complete works (1560) the poems plot their relation to space and place in his delightfully "amorous universe." Slowly the cosmographic pretensions become attenuated and the geography of contingency more pronounced. At the end of the 1550s, more embattled than he had been, the poet looks outward and around himself to discern who and where he is. The period that stretches from the end of the second or *Nouvelle continuation des amours* (1556; Weber and Weber 1963/1998, 215–57) until the resurgence of the poet in his *Discours* seems, like the two volumes of *Meslanges,* of mixed and motley aspect. Historians have wondered whether the poet continued to have qualms about the reception of his writing, or even if his effort to immortalize himself in "complete works" at the age of forty might have been born of a fallow moment of indecision, critical reflection, and doubt about what to do next and, no less, about his own economic well-being. History might indicate why. Robert de la Marthonie deprived Ronsard of the Priory of Saint-Jean-de-Côle, which Henri II had recently accorded to the poet and his brother. Ronsard appealed to his protector, the cardinal of Lorraine, and even to Marguerite de Navarre before, in exchange for the lost priory, he obtained Saint-Lucien de-Warhie, a presbytery in the diocese of Beauvais, thanks in part to the intervention of his protector Odet de Châtillon.

But already in 1556 Ronsard was said to have felt abused. He avows disillusionment in a letter to Charles de Lorraine at the end of the *Second livre des hymnes.* His brother Claude dies on 30 September of the same year, and the legacy he leaves is subject to familial dispute. In 1557 his friend the connétable de France, Anne de Montmorency, is defeated in battle at Saint-Quentin at the hands of Emmanuel-Philibert de Savoie. As he reedits the two *Continuations* he leaves the curacy of Challes and is entitled as the cleric of Fontaines in the diocese of Le Mans. Early in 1558 he squabbles with Nicolas Thibault, the abbot of Saint-Calais, over the abbot's failure to pay annuities. Ronsard writes an exhortation to herald the king's battle at Amiens, and not long after, while following the court, he writes an exhortation for peace. Early in 1559, three months after the death of Mellin de Saint-Gelais, Ronsard becomes a counselor and chaplain to the king. On 23 February he is granted a privilege to launch his complete works, just before more politics intercede with the Treaty of Câteau-Cambresis, the occasion of which allows him to write another circumstantial piece in honor of the cardinal of Lorraine. On 22 June Elizabeth of France marries Philip II of Spain, and one week later Henri II is mortally wounded in a festive tournament in which Anne de Montmorency's lance strikes the king in his left eye. Ten days later, on June 30, Ronsard's patron meets his demise at the age of forty. Ronsard soon publishes his *Second livre des meslanges,* in which, in a preface expressing grief over the death of the king, he wonders if he will continue to write poetry. At the same time religious turmoil anticipating the Wars of Religion begins to spread. News of the strife of Fort Coligny, Nicolas de Villegagnon's colonial venture in Brazil, reaches French shores. On 1 January 1560, Joachim Du Bellay dies at his desk after sharing a dinner with his loyal friend. By the end of the year Ronsard publishes at Gabriel Buon, his new editor, the four volumes of his *Oeuvres complètes,* which include as many dedicatory poems.

These traits are sketched not to have biography determine how and why the verse changes. The facts attest to the situated and contingent nature of the new poems in which the "amorous universe" cedes to local turmoil. The protean magus in the *Amours* gives way to an existential figure anchored in the time and place of circumstance and contingency. If the poet coordinates his work with the mechanism of the celestial machine, as he had in the philosophical tenor of the *Hymnes* of the middle 1550s, like the *Amours,* it is because, like the poet himself, the time is out of joint. The "moment" of 1559, when the *Second livre des meslanges* is published, offers the first indication of the poet's becoming a precursor of Montaigne's ideal topographer: a person with an errant but discerning eye, who looks at the world from within and without, who accounts for the shift from a cosmic worldview to that of an observer of partial things and of the chancelike nature of events happening in his midst.

The *Oeuvres* reprint twenty circumstantial sonnets from the *Second livre de meslanges* and add material, some of which arches back to the environs of 1555–57, the years in which oceanic voyage and travel to the Holy Land were prominent. In 1554 André Thevet launched a sumptuous edition of his *Cosmographie de Levant,* in which, following Polydorus Vergilius, he argues for the authenticity of fact born of ocular experience. Pierre Belon du Mans was publishing or researching his narrative of travel and proto-natural histories of fish and fowl. In 1557 and 1558 Thevet's *Singularitez de la France antarctique,* one of the first great "breviaries" of ethnography, comes forward with requisite controversy. Ronsard engages the perils of travel much as Thevet: bookishly, in a study, through a daring mix of classical and modern experience, far from strident claim of an ocular encounter with alterity, and more in keeping with the *Odyssey* and the new cult of its hero as a world historical figure than a tried and tested traveler.[9] In his two copiously illustrated works Thevet follows a convention by attesting that the naked eye is the best witness of truth at the same time it instructs the reader to look at the new and strange worlds in detail.[10]

Crucial for the context of the *Meslanges* of 1559 are the two inaugural poems in which fortune, contingency, and oceanic travel open the work onto an ethnographic plane. The *Second livre* begins with an *Elegie* (later titled *Discours*) to "Monseigneur le Reverendissme Cardinal de Chastillon" [My lord the Most Reverend Cardinal of Châtillon] in which the goddess of fortune is associated with the unpredictable and mixed—*pesle-mesle,* a word refracting the title—character of life. The conditions of the court, he asserts, are difficult. Flatterers abound, and so also sycophants who follow on one's heels. Fate blows the way the winds bend the stalks of wheat in the month of May, to the left and right, forward and backward, sometimes with fury, and at others "comme un tourbillon qui chassé de tonnerre / Premier en limaçon vient baloyer la terre" (fol. 4v; L10, 13; PL 2, 799, lines 75–76) [Like a tornado, chased by thunder, / just as a snail happens to sweep the earth], the spirals of wind resembling the shell of the gastropod that moves at its own pace. The breeze of fortune requires us to arm ourselves with virtue. Better that persons live in the fields they till and by the forests they planted in their youth, which they can now behold in their later years. The rustic country life, honest and simple, is worth far more than the ups and downs in the retinues of princes in their palaces. Drawing on the topical distinction between the city and the country, in praising Châtillon for his fortitude in the public domain, Ronsard takes a position recalling that of Philerme of *Saulsaye:*

> Quant à moy, j'aime mieux ne manger que du pain,
> Et boire d'un ruisseau puisé dedans la main,
> Sauter ou m'endormir sur la belle verdure,
> Ou composer des vers pres d'une eau qui murmure.
> Voir les Muses baler dans un antre de nuit,
> Ouir au soir bien tard pesle-mesle le bruit
> Des boeufs & des agneaux qui reviennent de paistre:
> Et bref, j'ayme trop mieux ceste vie champestre;
> Semer, enter, planter, franc d'usure et d'esmoy,
> Que me vendre moy-mesme au service d'un Roy.
> (fol. 7r; L 10, 14: lines 205–14; PL 2, 801, lines 153–62)

> [As for me, I prefer only to eat bread
> And drink water in my hands drawn from a brook,
> Gambol or nap on lush verdure,

Or compose verse near murmuring waters,
See the muses dance in a midnight lair
Hear the noise late in the evening, pell-mell,
Of steers and sheep that return from pasture:
In sum, I love much more this country life
Of sowing, grafting, and planting, free of usury,
Than to sell myself in the service of the King.]

In their sheen the lines betray a doubt about the fortune of being where one is. Elsewhere Ronsard compared a poem to a tree and poetry to a forest: if the metaphor shares the temper of this context, the man who sees the "great forests planted in his youth" would be the poet who beholds an opus soon to be published in difficult circumstances. The retreat into the forest is a retreat into poetry, where the sounds of another nature inspire the labor of writing: in the act of sowing, grafting, and planting, much as he had with the trees of his recent youth. And the retreat cannot be quite into the world of the *Amours* that in 1559 would be far from the current state of things.

The tenor of the inaugural poem inflects that of the longer *Complainte contre Fortune* (soon also to be called a *Discours*) addressed to the same protector. Counting 456 lines in its first version, the poem praises Châtillon in light of the fickleness of fortune: "ce monstre cruel, hydeux, & plein d'effroy" (fol. 8v; L 10, 18, line 55; PL 2, 771, line 44) [this cruel monster, hideous, full of fright] is far from the Proteus whom he admired at the outset of the first *Continuation* of 1555. Destiny is good and bad, mixed, *mélangé,* such that he now both *historicizes* and *mobilizes* identical depictions of his solitary idylls in the forests far from the court. Shards of the previous descriptions now acquire a different luster. Ronsard tells Châtillon about how he had been prior to living under his good graces:

> Avant que d'estre à vous, je vivois sans esmoy,
> Maintenant sur les eaux, maintenant à recoy
> Dedans un bois secret, maintenant par les prées
> J'errois, le nourrisson de neuf Muses sacrées.
> Il n'y avoit rocher qui ne me fust ouvert,
> Ny antre qui ne fust à mon oeil descouvert,
> Ny belles sources d'eau que des mains ne puisasse,
> Ni si basse vallée où tout seul je n'allasse.
> (fol. 9r; L 10, 19–20, lines 79–86; PL 2, 772, lines 67–74)

[Before seeing you I lived without fret,
Now on the waters, now in repose,
Now in a secret forest, now along the fields I went errant,

> Nursed by the nine sacred muses.
> No stone was left unturned,
> Nor was there lair that my eye failed to discover,
> Or beauteous spring in which I could plunge my hands,
> Or a valley too deep where alone I might go.]

All the movement and force of the *Amours,* witnessed in "Ciel, air, & ventz . . ." and "Comme un chevreuil . . ." are reproduced here ("now . . . now . . . now"), but in the indefinite past. He and his eye wandered in happy abandon when he drew inspiration from a real world. The landscape, at once conventional and bearing signs of an autobiographical moment, quickly becomes the setting for an inventory of his poetic worth. All of a sudden the poem becomes a résumé. Documenting his poetic mettle, he notes,

> Il n'avoit François, tant fust il bien apris,
> Qui n'honorast mes chants & qui n'en fust espris:
> Car tous ceux que la France en ce sçavoir estime,
> S'ils ne portent une envyeuse lime,
> Justes confesseront (& dire je le puis)
> Qu'avecques grand travail tout le premier je suis
> De Grece cy conduit les Muses en la France,
> Et premier mesuré leurs pas à ma cadence,
> Et en lieu de langage & Romain & Gregeois
> Premier les fis parler le langage François,
> Tout hardi m'oposant à la tourbe ignorante.
> (fol. 9r–v; L 10, 20–21, lines 91–101; PL 2, 772, lines 79–89)[11]

> [There was not a Frenchman, however so learned,
> Who failed to honor my songs and be enamored of them:
> For all those that France esteems in this wisdom,
> If they do not carry an envious file
> Will justly avow (and I can say so)
> That with this great labor I am the very first
> To have let the Muses into France,
> And who first measured their steps to my cadence:
> And in place of languages Roman and Greek
> I was the first to make them speak French,
> And with pride I opposed myself to the ignorant rabble.]

A pastoral turns into a self-defense, which then turns into a self-colonial operation. He almost militantly welcomes a benign invasion in which linguistic cleansing will become a national priority. An allusion is made to why he felt he could replace Mellin de Saint-Gelais as the laureate poet, and an invective is launched against the curriculum of the Sorbonne, opposed to the study of classical Greek. The words indicate why he will be justified in presenting his *Oeuvres complètes* to the world about him. About his accomplishments Ronsard is limpidly clear. At the same time he discloses that he can no longer do in the same manner what he had done in the *Odes* and *Amours*. An irrevocable line of divide with the poet of the *Amours* is drawn when the performative voice of the present draws its energies from that of the past:

> *Je fis des mots nouveaux, je restauray les vieux,*
> *Bien peu me souciant du vulgaire envyeux . . .*
> (fol. 9v; L 10, 21–22, lines 107–8; PL 2, 772, lines 95–97)
>
> [I fashioned new words, I restored the old,
> Scarcely fretting over the envious crowd . . .]

The assessment leads the poet to wonder why he learned the pathway to the Louvre; in turn, he wonders how the goddess Fortuna might indeed be the austere force of the celestial and terrestrial machine; and why he ought to address himself to her, she who may have guided him in the creation of his "amorous universe," like a nourishing mother, to have him compete with God. She is synonymous of all or *Tout:* "Bref, *tous* les accidens de la terre & de l'onde, / Et *tout* ce qui tourmente ou resjouïst le monde" (fol. 11v; L 10, 27, lines 222–23; PL 2, 775, lines 209–10; emphasis added). [In short, all the accidents of the land and the waters / And all that brings torment or joy to the world.] If the world is *pesle-mesle,* if it is a *grand meslange* and no longer an infinite variety of things, it is because of her astral force. The poet recounts how she sent a demonic intermediary, Malheur ("Misfortune") to inhabit the mind of the "Vandomois Ronsard" to cause untold mischief during his residence in Paris. After his misadventures he wonders why he had looked in vain for Hope to find the favor he needed, and for all that he often thinks of traveling elsewhere, to new or other worlds.

Antarctic France

At this point in the *Complainte,* the self-defending poet uses the topos of flight to engage political counsel about colonial policy. Ronsard embodies the character of the *topographe* whom Montaigne will praise twenty-one years later. The recent publication of Thevet's *Singularitez* motivates, it seems, the fantasy of going "anywhere out of this world." Now the issue concerns less the fantasy of entering into a classical landscape as he had in "Les isles fortunées" in 1553 (appended to the second edition of the *Amours* and addressed to its learned commentator, Marc-Antoine de Muret) than a sense of immanence and of human fragility. The persona of Fortune blends into the atomic substance of the things he describes, especially where he praises Châtillon for having helped to underwrite Villegagnon's colonial expedition to Guanabara:

> *Je veux aucunesfois abandonner ce monde,*
> *Et hazarder ma vie aux fortunes de l'onde:*
> *Pour arriver au bord, auquel Villegagnon*
> *Sous le pole Antarctique a semé vostre nom.*
> (fol. 14r; L 10, 33, lines 345–48; PL 2, 777, lines 317–20)

> [Sometimes I want to leave this world
> And hazard my life on the fortunes of the waters:
> To land on the shore where Villegagnon
> Planted your name under the Antarctic Pole.]

Risking fortune on the high seas (as Corrozet's nautical emblems had attested) is so treacherous that the goddess of the same name would have to accompany him to the ends of the known world. Above and beyond depicting himself as merely an armchair traveler (hence Ronsard seems to be a French anthropologist, a structuralist *avant la lettre*), the poet reflects on the nature of encounter and of the contact that Europeans have made with indigenous peoples of South America. Villegagnon, whom he addresses directly (L 10, 33, line 353; PL 2, 778, line 325), is chastised for wishing to civilize (*rendre fine,* which can be read as Cotgrave defines it, to make "wittie, craftie, subtile, cunning, wilie, fraudulent, cautelous, beguiling; also fine, small, prettie, curious; perfect, exact, pure, exceeding good, of the verie best; also most, very utmost") an unknown people that "*erre innocentement tout farouche & tout nu*" (fol. 14r; L 10, 34, line 356; PL 2, 778, line 328; emphasis added) [wanders innocently, naturally, ferociously and entirely nude], nude of—but not without—clothing and malice, who has no cognizance of the names of virtue, vice, of the senate or the king, "who lives in its pleasure, / borne by the appetites of its first desire." The initial traits of the portrait resemble those that Ronsard had drawn of himself when he lived in his sylvan world and where the forests he frequented had inspired the creative drive of his poetry. Thus, not at all in the topos of the noble savage, and without reference to classical models, Ronsard describes a nature that lives in total innocence with itself, and that by implicit contrast with the French court is entirely egalitarian in its political economy:

Et qui n'a dedans l'ame, ainsi que nous, emprainte
La frayeur de la loy, qui nous fait vivre en crainte:
Mais suivant sa nature est seul maistre de soy:
Soymesmes est sa loy son Senat, & son Roy:
Qui a grands coups de soc la terre n'importune,
Laquelle comme l'air à chacun est commune,
Et comme l'eau d'un fleuve, est commun tout leur bien,
Sans procez engendrer de ce mot Tien, & Mien.
(fol. 14r; L 10, 34, lines 361–68; PL 2, 778, lines 333–40)

[And who has not within the soul, as do we, imprinted
Fright at the law that causes us to live in fear:
But following one's nature is to be master of oneself:
One's being is one's law, his or her Senator and King:
Who with great thrusts of the hoe does not bother the earth,
Which, like air, is common to everyone,
And like the water of a river, is their common wealth,
Without engendering conflict from these words: Yours, and Mine.]

The description, allusively directed against the litigations over property in the "other" world *over here,* in Ronsard's milieu, is less idealizing than neutralizing in respect to concurrent representations of the savage or cannibal. Air, earth, and water are common to one and all.

But what of fire, the fourth of the elements needed to complete the description? Absent from the enumeration, the element is later found in the political dimension of Ronsard's reflections on alterity. He tells Villegagnon to leave the Indians alone and begs him not to break the "tranquil repose" of their "first life." If he were to have their lands surveyed and divided, striated and crisscrossed with lines in the manner of a topographer's survey, their ways of life would be destroyed: Western mores of measure and of governance need not be imposed because of what we witness in the "old" world:

Las! Si tu leur aprens à limiter la terre,
Pour agrandir leurs champs, ils feront la guerre,
Les proces auront lieu, l'amitié defaudra,
Et l'aspre ambition tourmenter les viendra,
Comme elle fait icy nous autres pauvres hommes,
Qui par trop de raison trop miserables sommes.
Ils vivent maintenant en leur age doré.
(fol. 14r; L 10, 34, lines 373–79; PL 2, 778, lines 345–51)

[Alas! If you teach them how to limit their land
They will then wage war, to augment their fields.
Litigation will follow, friendship will be defrauded,

And bitter ambition will come to torment them,
As it does here among our own impoverished people
Who by surfeit of reason are now in misery.
Today they live in their golden age.]

The past that is their present—*maintenant*—is ideal only by contrast
with the dystopic condition of life on this side of the hemisphere. If it
is an age of "gold," it is tempered by the figure in which it is cast. Gold
bears on the element of fire shown in the evocation of the ages of man
in which, in a dazzling passage, the forge and the furnace recall the
effects of *iron*. Ronsard insists that to "civilize" the Indians *(les rendre
trop fins)*, Villegagnon will see them set fire to the colony established on
what he and his men called "Fort Coligny" at Guanabara:

> *Certes, pour le loyer d'avoir tant labouré,*
> *De les rendre trop fins, quand ils auront l'usage*
> *De cognoistre le mal, ils viendront au rivage*
> *Où ton Camp est assis, & en te maudissant,*
> *Iront avec le fer ta faute punissant,*
> *Abominant le jour que ta voile première*
> *Blanchist sur le sablon de leur rive estrangere.*
> *Pource laisse les là, & n'atache à leur col*
> *Le joug de servitude, ainçois le dur licol*
> *Qui les estrangleroit sous l'audace cruelle*
> *D'un Tyran, ou d'un Juge, ou d'une loy nouvelle.*
> (fol. 14v; L 10, 35, lines 380–90; PL 2, 778, lines 352–62)

[Surely for the wages of having labored so much
To civilize them, when they will have known
How to recognize evil they will come to the shore
Where your camp is placed, and in cursing you
They will relentlessly punish you for your fault,
Abominating the day that your first sail
Whitened on the sand of their foreign shore.
Thus please leave them there, and do not pillory their neck
With the yoke of servitude, or the hard iron
That would strangle them under the cruel audacity
Of a Tyrant, of a Judge, or of a new law.]

A critique of colonial enterprise could not be clearer. Introduction of
Western technology will upset a steady state or "cold" society.[12] The frail
setting of the colony (on the island "où le camp est assis") will be subject
to siege. When they "go with fire" to settle accounts with the settler, the
effects of the age of iron will replace the setting of the age of gold. Force
of the critique intensifies in the temporality of the same lines. The "day
that your first sail whitened . . ." has a bleaching effect on the shores of

the "foreign border." No sooner is the conqueror's white shadow cast than *the land becomes foreign to the inhabitants themselves.* Servitude is shown imposed upon the natives in the very fashion that Étienne de La Boëtie will soon put forward in his *De la servitude volontaire* (1572). With requisite distance it can be said that in view of Claude Lévi-Strauss's reflections on the political order of the Nambikwara and Tupinamba, Ronsard's intuition about the nefarious effect of a "new law" has uncanny validity.[13] The author of *Tristes tropiques* reflected on political economy no doubt through the filter of La Boëtie and Montaigne, to the effect that an indigenous leader gains the approbation of his people by deed, example, and generosity. The social compact is held together through voluntary servitude, that is, an "unyoked" contract in which goodwill—*volonté*—is granted to the leader. Here the yoke makes the indigenous peoples strange to themselves. Their shores having become foreign to them, they are dispossessed and on the verge of falling into the murderous regime of Western mercantilism.

The passing remark about the New World brings perspective to the political order at a moment prior to the outbreak of the Wars of Religion. They will recur in the polemical verse of the early 1560s that, like the name given to this poem in its third edition (1567), is called *Discours*. Retrospectively the *Complainte* belongs to the author's political stirrings, said to lead to a Royalist and highly conservative position. Yet, at this moment, it is quite other: generous in spirit, cognizant of the deeds that informed its creation, and exemplary in the new style that turns a page after the foundation of the "amorous universe" of the early and middle 1550s. Ronsard is said to be the *poète engagé* after 1562, at the time of the tumult of the Conjuration of Amboise that sparked the wars to follow over the next thirty-six years. Here Ronsard is not yet a great or staunch defender of the Catholic cause. He offers counsel on the state of things at home, in his own condition, and in the world at large. He uses the mode of a complaint to craft a performative style by which a defensive posture attaches him to the country and the landscapes that had defined his amorous poetry.[14] The model of the poet-ethnographer is glimpsed in a few lines that acquire greater resonance both in the *Discours* of 1562 and 1563 and in their implied presence in the *Essais* of Montaigne. As of the *Second livre des meslanges,* Ronsard finds himself embroiled in a conflict in which his sense of geography is altered. By way of a reading of Montaigne it remains to see how and why, too, poets and essayists become ethnographers and topographers.

6

Montaigne and His Swallows

In "De la description d'aulcune region en plain, la latitude, longitude
& distance estans incogneues" (On the description of any region on
a flat plane where latitude, longitude and distance are unknown), the
first chapter of his appendix to Pieter Apian's *Cosmographie, Livret de
la raison, & maniere d'escrire les lieux, & d'iceulx trouver les distances
oncques paravant veu* (Book of the method and manner of plotting
places, and from them finding distances never before seen), Gemma
Frisius describes what it takes to make an informed measurement of
where one is.[1] Both geometry and a sense of place are key. To draw a
map of

> une entiere contrée, cherchez, premierement d'une ville dont il vous plaist com-
> mencer, les situations de tous les lieux gisans, & icelles pourtraiez au plain,
> premierement descripvant un cercle d'un point mis à vostre volunté. . . . Et pour
> éviter beaucoup de peine, montez sur la plus haute tour de la ville comme par un
> beffroy regardez à l'entour. Apres allez à une autre ville, & la faictes semblable-
> ment avec les angles de position de tous les lieux circonvoisins.[2]

> [an entire countryside, seek, first, from a city whence you wish to begin the
> situations of all the standing places, and portray them on a flat plane, initially by
> describing a circle about any axis placed as you wish. . . . And to avoid difficulty,
> climb to the top of the highest tower of the city, such as a belfry, and look all
> about. Then go to another city, and do identically with the angles of position of
> all the neighboring places.]

The topographer is asked to choose a town—any will do—and to look about and around it to get a sense of the "situation" prior to completing a trigonometric survey. He uses his astrolabe to measure the altitude of the tower from different points of view and, along the way, he disrupts the birds that fly out from under the rafters (Figure 37).

By climbing the tower or looking about from a belfry, the surveyor will avoid unneeded difficulty (*peine* in French) or travel (*peregrinationem* in the Latin edition). From the belfry the landscape can be seen and appreciated from all angles. The reader wonders if indeed Michel de Montaigne would have sensed a similar art of description from the third floor of his own tower. Would he, by looking at other notable places on the horizon, have identified himself in direct relation to the place where he was wont to write? Ever since Sainte-Beuve, who imagined Montaigne in his "ivory tower" (both the *tour d'ivoire* and the tower whence to see, *la tour d'y voir*), the modern reader of the late essays wonders what really was the experience of writing from the inner space of his library on the upper floor.[3] Or, too, the reader might speculate that the essayist, the harried homeowner of "De la vanité," climbed to the crenellations in late fall to look about as he cleared the gutters of leaves and debris.[4] Then, in the annual task of spring-cleaning he would have noted how the swallows, having reappeared on the horizon, were building their nests under the eaves or in the nooks and crannies of his chimneys. Or so he says, in reflecting on the frailty of human reason in the dense pages of the "Apologie de Raimond Sebond," when wondering if the songbirds surpass humans in their sense of space and place:

> Les arondelles, que nous voyons au retour du printemps fureter tous les coins de nos maisons, cherchent elles sans jugement et choisissent elles sans discretion, de mille places, celle qui leur est la plus commode à se loger? Et, en cette belle et admirable contexture de leurs bastimens, les oiseaux peuvent ils se servir plustost d'une figure quarrée que de la ronde, d'un angle obtus que d'un angle droit, sans en sçavoir les conditions et les effets? Prennent-ils tantost de l'eau, tantost de l'argile, sans juger que la dureté s'amollit en l'humectant? Planchent-ils de mousse leur palais, ou de duvet, sans prevoir que les membres tendres de leurs petits y seront plus mollement et plus à l'aise? Se couvrent-ils du vent pluvieux, et plantent leur loge à l'Orient, sans connoistre les conditions differentes de ces vents et considerer que l'un leur est plus salutaire que l'autre? (432/501)[5]

> [Are the swallows that we see upon the return of spring, ferreting about every cranny of our houses, looking aimlessly and choosing without discretion, from a thousand possible places, the most accommodating in which to be lodged? And in this beautiful and admirable contexture of their buildings can the birds

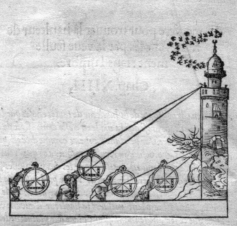

I L aduient fouuent qu'on ne peult aifement trouuer la diftāce d'aucune chofe,a caufe de quelque empefchemēt, En ce cas il fault par aultre
moyē befongner. Et au premier il fault fçauoir reduyre les parties de vm
bra verfa,les parties de ymbra'recta,ce que faict par telle artifice.Multi
pliez 12.en foy mefmes, & font 144. lefquelz partez par les parties de
vmbra verfa,lefquelles tu trouuéras. Et les parties ou degr.de vmbra ver
fa,ferōt couertis on degrez d'ymbra recta. Or dōc s'il te plaift fçauoir la
haulteur d'aucune chofe,tenez vo° en quelque lieu planier,et pēdez l'an-
neau par le fillet,cōme dict eft icy deuāt,changeāt le fillet de ça & la,iuf
ques a ce que vous regardez droict par les deux pīnules le fōmet de la cho
fe,et notez les parties du fillet,et fignez auffi la place d'une marque. A-
pres aprochez ou reculez felō la cōmodité de la place,d'autātqu'il plaift,
mais que ce foit la droicte voye vers la chofe qui eft a mefurer. Et de rechief regardez comme par auant par les deux pinnules. Adōc fi en quelque fois le fillet vous monftre parties de l'umbre renuerfée, il les fault reduyre aux parties d'umbra recta, par la maniere qu'auons dicte. Apres
mefurez la diftance qui eft entre les deux lieux de voftre ftation. Puis o-
ftez les parties moindres des plus grandes en nombre,le refte gardez pour
le diuifeur. En apres multipliez la diftāce par 12. & le nombre produict
partez par le diuifeur gardé, & le nombre qui en prouiendra fans doub-
te, vous monftrera la haulteur de la chofe qui eftoit a mefurer de voftre
œyl en hault. Comme par exemple, Pofe que le fillet au premier lieu mō-
ftre 8. parties d'umbra droicte. Et le fecond lieu 9.parties de L'umbre rē-

uer-

Figure 37. Gemma Frisius, treatise on measurement of local places in Apian, *Cosmographie* (1551), fol. 73r. Courtesy of the James Ford Bell Library, University of Minnesota.

make use rather of a square figure than one that is round, of an obtuse rather than a right angle, without being aware of their conditions and effects? Do they now take water, and then clay without judging that the hardness is softened by being moistened? Do they furnish the floor of their palace with moss or down without foreseeing that the tender members of their little ones will live within more softly and with greater ease? Do they protect themselves from the rainy wind and situate their lodging to the East, without being aware of the different conditions of these winds or considering that the one is healthier for them than the other?]

With the mention of the *arondelles,* Montaigne, in praising animals at the expense of humans, sets in motion a rhapsody of natural history that takes up a good part of the first half of the "Apologie."

When Montaigne looks at the swallows he revives the memory of his task of translating Raimond Sebond's (or Raimondus Sebundus's) *Theologia naturalis* to honor his father, Pierre Eyquem, the rustic builder of the estate who welcomed, notes Montaigne at the very beginning of the "Apologie," the presence of learned men in his milieu.[6] Completed in 1569, the essayist's labors yielded a second edition in 1581, the year following the publication of the first two books of the *Essais* at Simon Millanges in Bordeaux. Inverse to Sebond, who had established the medieval world picture and chain of hierarchies stretching from inert forms to animals and then from humans to God, Montaigne thrusts man back to the bottom of creation. He crushes human presumption and vanity through appeal to every type of reason that had been collected in Sebond's treatise. An apologia for Sebond is advanced and no sooner inverted than, all of a sudden, the rhetorical machinery of the essay seems to lose control of itself. The memory of Sebond is brought forward as if, it soon avers, the cosmic diagram of the *Théologie naturelle* were a piece of bait set before an enraged beast, the essayist himself, who exhorts endless admonishment in order to dash every illusion we have about the goodness of science and belief. The result becomes the most dominating and voluminous chapter of all the *Essais.*

A Form of Content

Critics in the line of Pierre Villey have interpreted the monstrous essay to show that it stands as a crucial but intermediate episode in a writer-philosopher's career that espouses Stoic belief, undergoes a fideistic crisis marked by an excess of "pyrrhonizing," and that finally leads to the Epicurean and Socratian pleasure of wisdom attained after trial and essay.[7] Despite the appeal offered by the image of a subject working through a dilemma of doubt, the almost "introjective" volume and mass of the "Apologie"—introjective because it holds and suspends all knowledge within its own paginal space—stages in simultaneously geometrical and narrative ways a compelling crisis.[8] It opens a gap between its review of inherited constructions of space and the stunning creation of its own form. The text puts forward a mix of the discourses of experience, received schemes of *habitus,* and the major modes of mapping the world at large in what seems to be of monstrous proportion in relation to the surrounding essays of more modest measure. It is cosmographic in size but topographical in the attention it pays to detail and in the way it produces, as the author had stated in "Des cannibales," "particular accounts" of the intellectual places he has known.

In its inversion of Sebond's world picture the essay betrays a broader concern with what happens when a subject discovers that he or she has *already* been plotted and located by spatial schemes that for strategic ends have summarized God's creation. The latter are called into question for dictating the way homily and inherited truth build illusions of human control exerted upon a space bequeathed by God. The conflicting orders of knowledge that inhabit the subject inspire a micropractice or topography (a personal essay) that translates the plan of a self-enclosing macrospace (a world, even a textual cosmos, whose limits are defined by the shape of the text that can extend almost infinitely) that swallows and reconfigures inherited cosmic diagrams. In a coextensively discursive and visual order, in what contemporary philosophers would call its form of content, the essay also stages a crisis in which infinite space and duration overwhelm any control that cosmography might exert upon human knowledge and the illusion of power that goes with it.[9]

At the same time and to the contrary, the "Apologie" also demarcates the limits of a finite world beyond which there lies an absolutely unthinkable area of nothing, a *dehors,* that cannot be reduced to or approximated by language, reason, or any schematic system.[10] In negotiating this paradox the essay summons its own attempt both to chart and to account for, printed book that it is, a two-dimensional view of the state of knowledge in its time. The essay can thus be called a map of paradox in which are specified some of the ways that subjectivity and spatial consciousness are related. In order to see how *revenons à nos hirondelles,* let's get back to our swallows. In the early pages Montaigne invokes the presence of swallows just after he has inverted the great chain of being:

> La presomption est nostre maladie naturelle et originelle. La plus calamiteuse et fraile de toutes les creatures, c'est l'homme, et quant et quant la plus orgueilleuse. Elle se sent et se voit logée icy, parmy la bourbe et le fient du monde, attachée et clouée à la pire, plus morte et croupie partie de l'univers, au dernier estage du logis et le plus esloigné de la voute celeste, avec les animaux de la pire condition des trois; et se va plantant par imagination au dessus du cercle de la Lune et ramenant le ciel soubs ses pieds. (429/497–98)

> [Presumption is our natural and originary malady. The most calamitous and frail of all creatures is man, and now and again the most boastful. It (the creature) senses and sees itself lodged here, in the muck and dung of the world, deadest and most stagnant part of the universe, at the last floor of the lodge and the most remote from the celestial vault, with the animals of the worst condition of the three; it goes about planting itself in its imagination above the circle of the Moon and bringing the sky beneath its feet.]

By isolating himself from animals the human creature loses all cardinal and corporal bearing; it becomes oblivious to the inner space inhabited and articulated by the bestial world, that is, "les branles internes et secrets des animaux" (ibid.) [the animals' inner and secret stirrings]. Citing evidence from cosmographers' reports about the surface of the earth, Montaigne shows that if certain nations interpret canine gestures to be forms of divination, it follows that, because of an implicit parity between man and animal, unspoken signs, notably (1) the language of the deaf, (2) the glances that lovers furtively and secretly share, and (3) hand gestures, ineffably convey what Montaigne proceeds to enumerate in a monstrous volley of verbs. Without speaking, "nous requerons, nous promettons, appellons, congedions, menaçons, prions, supplions, nions, refusons, interrogeons, admirons, nombrons, confessons, repentons, craignons, vergoignons, doubtons, instruisons, commandons, incitons . . ." [we request, we promise, call, dismiss, menace, pray, beg, deny, refuse, interrogate, admire, enumerate, confess, repent, fear, shame, doubt, instruct, order, incite] and so forth, with twenty-five more instances, until the series is momentarily ended with "escrions, taisons" (454/500) [cry out, silence ourselves]. For a moment the text either goes wild or betrays the mechanism of a machine on the verge of going out of control.[11]

When it serves to prove the point that the essay will later claim about the utter lack of any communication that man holds with being, the *arondelle* swoops into the space of the discussion with such ease that the effect left by the list of verbs suggests how much beasts, more than and to the detriment of man, are in harmony with their milieus. Implied in the enumeration is a continuation and inversion of a medieval tradition, in which birds had been configured as volatile machines, as creatures of a herky-jerky demeanor that embodies mechanistic process in general.[12] But the swallows migrate through the text smoothly and even furtively. They burrow into an unsettling space that marks a more general and charmingly monstrous habitat. "Les arondelles au retour du printemps que nous voyons *fureter* tous les coins de nos maisons" are likened to four-legged, quasi-domesticated creatures, *ferrets,* which Conrad Gesner had typified, because of the etymons *furo, furetus,* and *furectus,* as weasels that "preyeth upon conies in their holes, and liveth upon stealth" (in Topsell 1967, 170).

In drawing attention to the difference between two species by confusing orders of two-legged (feathered) and four-legged (furry) beasts, the text also contrasts the bird to the ferret on the basis of architectural style and the construction of domestic space. One animal is implied to be a benign busybody, while the other generally signifies a ferocious burrowing beast. In the mix of the subject (swallow) and predicate (ferret) of the sentence, a monstrous form is born. Where the swallow is revered as a harbinger of life and of the return of spring, the ferret, as Gesner's editor notes, is wild and vicious. It can be trained and "tamed to hunt conies out of the earth. It is a bold and audacious beast, enemy to all other, except his owne kind, drinking and fucking the bloud of the Beast it biteth, but eateth not the flesh" (171). But the bird is an architect that does not burrow *into* and extract but builds *from* mud and stray matter. It knows how to solve the quadrature of the circle. As a form that is as round as its name, it inhabits the square order of the *coin* where it builds its lodging.

Swallows compass their nests without recourse to Euclidean reason, yet the habitat they construct is in harmony with the conditions and effects of the edges and nooks to which they apply mixtures of clay and water. The outer surface of the nest that Montaigne describes also displays an inner wealth and warmth. The muddy palace, furnished with moss and down, is warmed by exposure to the rising sun and protected from the winds from the North. The text exudes a uterine fantasy where, like a chick, its author basks in a warmth and plenitude enveloping his own members. His limbs are implied to resemble those of the birds where, in the nest, "les membres tendres de leurs petits y seront plus mollement et plus à l'aise" [the tender members of their little ones will live within more softly and with greater ease].

The vocabulary and its analogies become so confused that at least four idiolects are simultaneously registered in the description of the bird. The *arondelle* is an architect and a geometer; a practical scientist who knows how to build with nature's cement and stucco; a creature adequate and identical to its name, *ronde,* that produces an identical volume of infinitely intimate warmth; a volatile that flies above the earth but shares traits with burrowing or subterranean creatures. Finally, since the bird is commonly seen *au retour du printemps,* it is perpetually surrounded by the season of growth that it carries with it. Its name literally maps out the habitat that perpetually surrounds it. The orthography recalls that of the "aronde," the bird that is round, yet one embellished with wings, suggestive of a latent rebus of an *aronde-aile.* The swallow of the same form appeared in Jean de Meung's *Roman de la rose* and Clément Marot's *rondeaux,* displaying in its graphic shape the very pattern of its nomadic ways.[13] Most important, "les arondelles, que nous voyons au retour du printemps" are never described as migrating to a specific place *outside* the ken of the observer. In a fashion that conjugates the swallows' habits as they had been conveyed in late-medieval writing, the description implies that the birds merely come and leave. The bird tells us that what comes around goes around.

The swallow is not associated with a geography of migration to North Africa in the winter months and return to the Bordelais and Gascony in the spring. Arguably the space implied by the perpetual circularity of the bird's presence (it goes away, but we need not know where; it comes back, it builds, it breeds, but for the rest we don't care . . .) suggests that when it is not here it does not disappear: it is simply "around," in a space "off" or outside. The description crystallizes the reflections about cosmic space and *habitus* in the "Apologie": when tracked along a geodetically determined trajectory, the bird's migratory patterns yield a topography. Knowledge of the itinerary would put at a further remove a cosmic cause *outside* the effect of its own way of living. Were the bird used for the mapping of Mediterranean space as geographers, travelers, and natural historians had already done, it would serve human reason and experience. Yet even if the birds served as important measurements of physical space, they continued to be taken as signs of divine presence and seasonal change.

Belon's Birds

The history of the swallow in early books of ornithology informs these questions. In the work of the great natural historian of the age, Pierre Belon du Mans, Montaigne's casual observation about the *arondelle* is given different inflection. The last three chapters of his *L'histoire de la nature des oyseaux* (1555) describe and illustrate four species of swallow, all of which are "vulgaires aux paisans, villageois, et bourgeois de France" (376) [common to village peasants and the city dwellers of France]. "La petite hirondelle" (Figure 38) is thus described:

> Ses jambes sonts courtes, & les pieds faitz à la maniere des oyseaux qui se perchent. Lon pense qu'elle face ses petits deux fois l'an. Qui nous semble estre vraysemblable: car nous voyons qu'elle est absente au tant de temps hors de nostre païs, comme presente. Et pource qu'elle retourne lors que l'esclaire est en fleur, les autheurs ont donné le nom d'Hirondelle à l'esclaire, la nommants *Chelidonium*. Et tout ainsi que ceste Chelidoine a vertu de guerir les yeux, aussi pense lon que les petits de l'Hirondelle, aveuglez de la fumee des cheminees soyent gueris par l'herbe que la mere leur apporte dedens le nid. (fol. Lii r)

> [Its legs are short, and its feet made in the manner of perching birds. It is thought that it hatches its young twice per year. Which is likely, because we see it as much absent from our country as present. And because it returns when the *esclaire* flowers, authors have named it *Chelidonium*. And just as this Chelidoine has the virtue of curing eyes, it is thought, too, that the Swallow's chicks, blinded by the smoke of our chimneys, are cured by the herb that the mother brings to them in the nest.]

Present as much as absent (and present here, in description and emblematic illustration in both Belon's compendium and the "Apologie"), it is always about and around. In the physiological sense, too, it exemplifies what Gregory Bateson calls the "loop structure" in "feedback" relations. It knows how to adjust to human pollution through an intelligent dietary regime that it passes to its chicks that would be blinded by the smoke emanating from chimneys adjacent to its nests. For the naturalist Belon, like Montaigne, the *hirondelle* is a composite creation, an organic bird-machine, that is armed with both "feet" and "claws," and its attributes are congruent with its total social fact: its season, its home, its food, and its habitus amount to its reason.

On at least two points Belon's description offers a striking contrast to Montaigne's casual observation. First, the natural historian, following Pliny, elsewhere notes that the bird is said to migrate to and from Africa and France. He begins to demystify the force of presence that the

Figure 38. Pierre Belon, the swallow *(hirondelle)* in *L'histoire de la nature des oyseaux* (1555), p. 379. [Typ 515.55.201 (F) B] Houghton Library, Harvard University.

swallow possessed in the medieval tradition to which the "Apologie" returns. In the encomium of birds that serves as a preface to the natural history, Belon remarks, in a way that seems to inspire Montaigne's own admiration of the *arondelle:*

> Qui croiroit que les hirondelles & autres petits oysillons, qui demeurent seulement [à] l'est en nostre Europe, puissent avoir si tost basty leurs nids, & avec si grande industrie? Il n'y a homme qui ne doive estre incité à son devoir par l'exemple de la diligence des oyseaux passagers, qui en moins de trois jours et nuicts ont passé l'Europe en Afrique. Qui leur apprend l'election des vents propices à cest effect, & choisir l'endroit du ciel pour l'eslever en l'aer, & ne faillir leur chemin sans guide, sinon nature? (fol. 111r)

[Who, unless it be nature, could believe that swallows and other little birds that live only in the east of our Europe can have so promptly built their nests with such industry? No one can fail to be moved in his labors by the example of the diligence of these passing birds that in less than three days and nights fly from Europe to Africa. Who teaches them to choose the winds that carry them and to select the spot in the sky to be carried aloft in the air and never stray from their course without a guide?]

The question that Belon poses hovers between presence and science. The swallow, an emblem of the beauty of God's creation, is also an implicit figure in the mapping of European space. Belon's orthography makes the essayist's *arondelle* look archaically poetic. He tends to detach the name from the thing itself, attenuating the "naturalistic" view of language that Montaigne revives with the bird a quarter of a century following the publication of *De la nature des oyseaux*.[14]

Second, and in a less immediate way, Belon's extensive praise of birds as analogies of man's character strikes the very theme that Montaigne uses to dislodge humans from the upper rungs of the ladder of being. In declining the virtue of plumed beasts the natural historian discovers that their attributes are related to writing, the very instrument of man's faculty of reason:

Les anciens, comme encor pour le jourdhuy les Grecs, Turcs, Arabes, Siriens, Perses, & touts autres hommes qui habitent en levant, n'ont aucun usage des plumes d'oyseaux pour se servir es leurs escriptures, comme nous faisons maintenant: mais ont des tuyaux de rouseaux ou cannes, qui est cause nous ne pouvons exprimer tel nom en Latin que le nommer *Calamus*. Car lon ne dira *Penna* pour parler d'une plume à escrire. Mais parlant comme Aristote aux livres *De natura & partibus animalium*, devons la tige ou caule [que] . . . les interpretes ont dit *Caulis in pennis*. (fol. 35–36)

[The ancients, as still today for Greeks, Turks, Arabs, Syrians, Persians, and all other men who live in the Orient, have no use for birds' feathers to be used for their writings as we do now: but they have straws of reeds or canes that tell us why we can only express them in Latin in the name of *Calamus*. For we would never say *Penna* to speak of a writing instrument. But as Aristotle states in *De natura & partibus animalium*, we have the quill or stalk that interpreters have called *Caulis in pennis*.]

Implied is that birds carry with them the technology that causes them to "write" the patterns of their presence into the world. Montaigne, by contrast, in the pages that debunk man's presumption, shows that for all of his creation (and it is inferred, his effects of writing), the human species remains the most *calamitous* and frail of all creatures on earth. The text of the "Apologie" is aware of the misnomer that Belon brings into view to bestow praise upon birds in general.

Whether Montaigne may have read Belon's treatise is moot. More important is that in the use of a theory of natural language generating the descriptions in both the "Apologie" and *De la nature des oyseaux,* the symmetry of the bird and its world in relation to man is reiterated, in particular where the human skeleton (Figure 39) is compared to the avian counterpart (Figure 40). The most distinguishing trait is the beak, which is described as what would be the analogue to the human's lips and teeth, and its unique collarbone that is a "lunette" (lens) or a "fourchette" (fork). The near-identity is made more compelling in both the spatial and the temporal registers of Montaigne's essay (temporal because assimilation of the narrative or itinerary of the text requires a long duration of time, and spatial for reasons adduced here).

In the rhapsody of beasts that takes up a first major movement of the "Apologie," the swallow returns on four more occasions, as according to the "seasons" that the slow reading of the essay requires. First, the bird is summoned to show that animals are well suited for their practice of venery, just as we hunt game: "nous allons à la chasse des bestes, ainsi vont les Tigres et les Lyons à la chasse des hommes; et ont un pareil exercice les unes sur les autres: les chiens sur les lievres, les brochets sur les tanches, les arondelles sur les cigales" (439/509) [we go off to hunt beasts, and thus off go Tigers and Lions to hunt man; and their practice with one another is similar: dogs for hares, pike for dace, swallows for cicadas]. Second, the swallow is lionized for an ecology that can teach human beings a good number of lessons about domestic space. In the third instance, beasts have taught us most of what we know, "comme l'araignée à tistre et à coudre, l'arondelle à bastir le cigne et le rossignol la musique" [like the spider for weaving and sewing, the swan and nightingale for music]. And, as Belon had already shown, "plusieurs animaux, par leur imitation, à faire la medecine" (442/512) [several animals, through their imitation to practice medicine]. Fourth, the birds have the gift of divination because, changing their household according to the seasons, they "montrent assez la cognoissance qu'elles ont de leur faculté divinatrice, et la mettent en usage" (448/518) [evince well the knowledge they have of their capacity to divine, and put it to use]. But most important, in a series of enumerations queued by the mechanical formula "quant à" ("as for," which almost endlessly quantifies)—"quant à la mesnagerie" (451/522), "quant à la guerre" (452/523), "quant à la fidelité" (454/526), "quant à la gratitude" (455/527), "quant à la societé" (457/529), "quant aux particuliers offices" (457/529), "quant à la magnaminité" (459/531), and others—the penultimate example of "cette equalité et correspondance de nous aux bestes" (460/532) [this equality and correspondence we share with beasts] invokes a fabled relative of the sparrow, the halcyon, to humiliate the human species further. Poets—Montaigne, too—recall how the world stopped and the sea calmed, seven days before and seven days after the winter solstice, when the bird nested at sea. Most remarkable was the nest it builds from various fish bones:

> qu'elle conjoint et lie ensemble, les entrelassant, les unes de long, les autres de travers, et adjoustant des courbes et des arrondissements, tellement qu'en fin elle en forme un vaisseau rond prest à voguer; puis, quand elle a parachevé de le construire, elle le porte au batement du flot marin, là où la mer, le battant tout doucement, luy enseigne à radouber ce qui n'est pas bien lié, et à mieux fortifier aux endroits où elle void que sa structure se desment et se lâche pour les coups de mer; et, au contraire, ce qui est bien joinct, le batement de la mer le vous estreint et vous le serre de sorte qu'il ne se peut ny rompre, ny dissoudre, ou endommager à coups de pierre ny de fer, si ce n'est à toute peine. Et ce qui plus est à admirer, c'est la proportion et figure de la concavité du dedans: car elle est composée et proportionnée de maniere qu'elle ne peut recevoir ny admettre autre chose que l'oiseau qui l'a bastie: car à toute autre chose elle est impenetrable, close, et fermee, tellement qu'il n'y peut rien entrer, non pas l'eau de la mer seulement. Voilà une description bien claire de ce bastiment et empruntée de bon lieu; toutesfois il me semble qu'elle ne nous esclaircit pas encore suffisamment la difficulté de cette architecture. (460/532)

> [that it conjoins and ties together, weaving them, some lengthwise, others cross-wise, and adjusting the curves and arched forms *(arrondissements)* such that in the end it forms a round vessel that can float errantly; then, when it has finished building it, it carries it to the surf where the sea, beating it quite softly, informs it about where to caulk what is not well tied and to fortify better the places where it sees that its structure is loose and releases it in the throes of the surf; and to the contrary, what is well joined, the beating of the sea tightens and strengthens it for you so that nothing will break, dissolve, or be damaged by the shock of stone or iron if it is not in vain. And what is most admirable is the proportion and figure of concavity on the inside; for it is composed and proportioned in a manner that it cannot receive and allow entry for anything other than the bird that has built it; for everything else it is impenetrable, closed, and hermetic; nothing can get inside, not even the water of the sea. Now there's a very clear description of this building taken from a good place; yet it seems to me that it does not inform us sufficiently of the difficult beauty of this architecture.]

Portraict de l'amas des os humains, mis en comparaiſon
de l'anatomie de ceux des oyſeaux, faiſant que les
lettres d'icelle ſe raporteront à ceſte cy, pour
faire apparoiſtre combien l'affinité eſt
grande des vns aux autres.

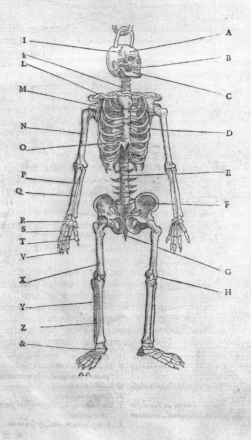

Figure 39. Pierre Belon, "Portrait of the mass of human bones, compared with the
anatomy of those of birds . . ." in *L'histoire de la nature des oyseaux* (1555), p. 40.
[Typ 515.55.201 (F) B] Houghton Library, Harvard University.

La comparaison du susdit portraict des os humains monstre combien cestuy cy qui est d'vn oyseau,en est prochain.

Portraict des os de l'oyseau.

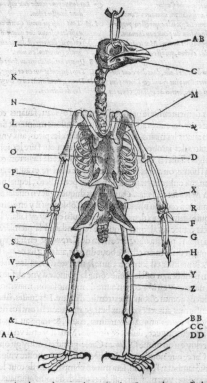

AB Les Oyseaux n'ont dents ne leures , mais ont le bec tranchant fort ou foible,plus ou moins selon l'affaire qu'ils ont eu à mettre en pieces ce dont ils viuent.

M Deux pallerons longs & estroicts, vn en chascun costé.

ᴣ L'os qu'on nommé la Lunette ou Fourchette n'est trouué en aucun autre animal , hors mis en l'oyseau.

D Six costes , attachees au coffre de l'estomach par deuât,& aux six vertebres du dos par derriere.

F Les deux os des hanches sont longs , car il n'y a aucunes vertebres au dessoubs des costes.

G Six osselets au cropion.

H La rouelle du genoil.

I Les sutures du test n'apparoissent gueres sinon qu'il soit boully.

k Douze vertebres au col, & six au dos.

d iii

Figure 40. Pierre Belon, "Comparison of the portrait of human bones shows how what here is of a bird is similar," in *L'histoire de la nature des oyseaux,* p. 41. [Typ 515.55.201 (F) B] Houghton Library, Harvard University.

Cribbed from Plutarch, the description is nestled into the text such that the implied writer, a *pescheur de rivage*—one of Belon's kingfishers— begins to inhabit the utopian cavity of the very nest it builds.[15] The fabled nest virtually "rhymes" with what everyone sees in the crannies and under the gutters of their houses, except that now the surrounding element is the sea. The uterine nest, folded into and protected by and from the mother sea, underscores the obsession that the "Apologie" everywhere makes manifest about lived and local space as well as lodg- ing and *habitus*.

At least two qualities can be deduced from the example of the halcyon and its placement in the encomium of beasts. First, the nest projects an abyssal image of a perfect architecture that floats within but also gives onto the world at large. It would not be unwarranted to beg a com- parison between the defensive structure of the "Apologie" as a whole and the figure of the author admiring the protective and impenetrable walls of the halcyon's nest. Floating at sea, as emblematists such as Gilles Corrozet had figured a similar nest (in the *Hécatomgraphie*), it is isolated but remains a self-protected space within the environing world (Figure 41). And at this point in the essay the bird's nest becomes the inverse and reflection of the massive verbal extension that surrounds it. Second, insofar as the figure is pictured adrift in an ocean with its creator ("l'oiseau qui l'a bastie"), the abode appeals to a cartographical motif that shapes much of the rest of the "Apologie." In exposing "man" as isolated in an infinite space of the world, Montaigne appeals to the *isolario* or book of islands, a genre that had been current from 1485 to 1573, as a model for constructions of relativity and ordered chaos. Two of the principal forms of the essay's content are therefore the nest and the island in an ocean of discourse. The isolario had allowed for a creation of taxonomies of things unknown and singular. As André Thevet had recently shown in the *Singularitez de la France antarctique* (1557) and in his more recent *Cosmographie universelle* (1575), a book- compendium could be conceived to identify verbal copia and fluvia at once alternately and simultaneously as objects and as flow, as islands, as flotsam, but also as process with severely defined limits. Until the constrictive sense of the world's perimeters had been apprehended by the 1580s, the island book could isolate information and extend analo- gies among the things contained within it, but all the while its form would be implied to exceed its own foliated space.[16] Folded within and at the center of the *Essais* is thus a form that extends beyond all that

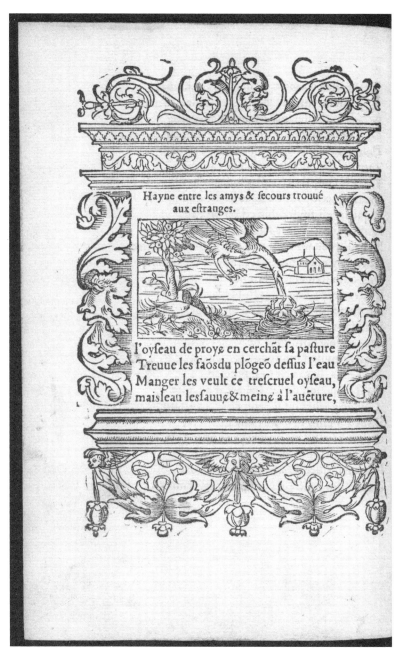

Figure 41. The *plongeon* or diving bird in Gilles Corrozet, *Hécatomgraphie,* emblem 30. [Typ 515.43.299] Houghton Library, Harvard University.

surrounds it, but in enveloping, swallowing, or introjecting the world as it does, it delineates a *nothing* outside it. Exhausting the space it inherits and reconstructs, the essay becomes a fabulously cosmographic event. In its graphic mass the essay fashions a finite world that marks, by virtue of the demonstration of its extension in respect to the other island chapters in the archipelago of essays, a closure that can both affirm and deny the presence of infinity. The passage of the birds in the "Apologie" indicates that the text sets out to fly over received phenomena so as to make them strange, but also to burrow into, work through, or, in the bestial mode, "ferret" them out from within. Essential to the displacement is the construction of a verbal extension in which cardinal points of reference are lost. Yet, at the point where the map of relativity, the isolario, seems to dominate the world picture of the "Apologie," its form gives way to a planispheric view of a closed and finished condition. Montaigne remarks:

> Ptolemeus, qui a esté un grand personnage, avoit estably les bornes de nostre monde; tous les philosophes anciens ont pensé en tenir la mesure, sauf quelques Isles escartées qui pouvoient eschapper à leur cognoissance: c'eust esté Pyrrhoniser, il y a mille ans, que de mettre en doute la science de la Cosmographie, et les opinions qui en estoient receuës d'un chacun; c'estoit heresie d'avouer des Antipodes: voilà de nostre siecle une grandeur infinie de terre ferme, non pas une isle ou une contrée particuliere, mais une partie esgale à peu près en grandeur à celle que nous cognoissions, qui vient d'estre descouverte. Les Geographes de ce temps ne faillent pas d'asseurer que meshuy que tout est trouvé et que tout est veu. (555–72/645)
>
> [Ptolemy, who was a great character, had established the limits of our world; all the ancient philosophers thought by establishing its compass, save a few remote islands that were beyond their ken: a thousand years ago a person would have pyrrhonized by calling into question the science of Cosmography and the opinions about it received by each and everyone; it had been a heresy to acknowledge the Antipodes: now we see in our century an infinite continental expanse, not an island or a particular country, but a part almost equal in size to what we used to know, that has just been discovered. The Geographers of our time do not fail to assure that as of now all is found and all is seen.]

In this celebrated passage, which juxtaposes cosmography and geography as well as "modern" and "antique" representations of space, in its errant and meandering way as it moves about in "rolling and flowing," the essay is arguing for a topographical—a close and detailed—reading of itself. No shard of knowledge, like a stone or pebble, goes unturned. Its immense volume in the context of the surrounding essays signals that

it is reacting to the fear of an end of things unknown or of terrae incog-
nitae. If it casts doubt on the veracity of the isolario as a representation
of an unlimited world, it follows that its shape is fittingly monstrous:
first, as a physical form whose mechanical and machinelike writing
(what a recent philosopher would call its spiritual automatism), swallows and digests all knowledge and all history.[17] Second, as a catastro-
phe folding the knowledge of the world into the field of its doubt, the
chapter uses its form as a place in which to lodge its imaginary author,
in a nest, on an island, in a labyrinth or any number of regions in order
not to be reduced to a monolithic "world picture."

Region and Religion

At this point, in guise of a conclusion, *revenons encore à nos hirondelles.*
The swallow never returns to the end of the essay to assure the order
of a seasonal continuity that would guarantee the order of God and
Christian faith—even if they are praised in the last sentences of the
essay. The bird nonetheless appears in an emblem of anamorphosis, as a
visual abstraction that extends further the monstrous figure it assumed
at once as a plumed volatile and a furry, four-legged beast simultaneously isolated from and in a symbiotic rapport with its surroundings. In
the final sentences the text reminds us that "il faudroit donc que nous
en fussions premierement d'accord avec les bestes" (598/676) [it would
be wise if we were first of all in agreement with beasts] before we begin
to think about our own relations with ourselves. To adduce the fluidity
of perception and the ever-changing condition of reason, Montaigne
appeals to inventions of wit that dissolve differences of signs and their
referents. Some of which are

[c]es bagues qui sont entaillées en forme de plumes, qu'on appelle en devise:
pennes sans fin, il n'y a oeil qui en puisse discerner la largeur et qui se sçeut def-
fendre de cette piperie, que d'un costé elles n'aillent en eslargissant, et s'apointant
et estressissant par l'autre, mesmes quand on les roule autour du doigt; toutes-
fois au maniement elles vous semblent equables en largeur et par tout pareilles.
(583/677)

[these rings that are incised in the shape of feathers that in a motto are called
endless plumes, there is not an eye that can discern their width or not be
taken in by this trick, that from one side they go in widening and in gathering
and narrowing from the other, even when they are turned around the finger;
nonetheless when they are handled they seem equal in width and everywhere
identical.]

On the mercurial surface of the words describing the ring we see in visual echo the "aronde ailes" of plumed rings that are both the swallows
and their migratory path of eternal return. The figure that continually
changes its proportion about the fingers is yoked to argue for the roll
and toss of a world in perpetual change.

The anticipation of the famous *branloire perenne* of "Du Repentir,"
the world in perennial stirring and tossing, comes as no surprise in an
essay celebrating the power of doubt. The bent and bending inflection
of both objects and languages in the final sentences recalls how swallows show the human species that it cannot obtain an ichnographic or
a bird's-eye view of the world in the way that "man," its presumptuous
cartographer, might wish:

> Finalement, il n'y a aucune constante existence, ny de nostre estre, ny de celuy
> des objects. Et nous, et nostre jugement, et toutes choses mortelles, vont *coulant
> et roulant* sans cesse. Ainsin il ne se peut establir rien de certain de l'un à l'autre,
> et le jugant et le jugé estans en continuelle mutation et branle. Nous n'avons
> aucune communication à l'estre. (586/679; emphasis added)

> [Finally, there is no constant existence, neither of our being nor that of objects.
> At once we, our judgment, and all mortal things go ceaselessly flowing and roll-
> ing. Thus nothing certain can be established from the one to the other, and the
> judging and the judged being in continual mutation and turmoil. We have no
> communication with being.]

The pairing of the rhymed participles is a reminder of birds cooing in
their nests, *roucoulant* in that, in the suggestive filigree of figures, as
organic machines the wheels of birds-as-a-world turn without friction
and in perpetual motion. The text fastens onto the figure and gives it a
cartographical valence in the sentences that lead to the end of "ce long
et ennuyeux discours . . ." (588/681) [this long and boring discourse].
Once the text reaches the greater historical remove of the Pre-Socratic
philosophers ("toute matiere est coulante et labile . . . jamais homme
n'estoit deux fois entré en mesme riviere . . ." [all matter is flowing and
smooth . . . never did a man swim twice in the same river . . ."], it can
roll, like the halcyon's abode, on the waves of the verbal flow that approximates what is being described, "avec la matiere coulante et flu-
ante tousjours" (588/680–81) [with matter always in flow and flux]. The
arondelle, now transmuted into the "labile" machinery of the essay's
own discourse and geography, retains within the verbal intervals of the
reiterated formulas the bestial and mechanical quality of the discourse
that plots the infinite and self-confined space of the "Apologie."

The passing observations on the beauty of swallows' nests in spring or the halycon's maritime domicile appear to confirm a dilemma seen in a widening gap between belief in received diagrams of space and the everyday, common, "micropractical" perception and tactile sense of things in flight and passage. But the textual process and the latent mapping of the essay allow its writer to become a bird, to build and nestle into an avian and aeronautical space; they also afford an errant eye on a world seen as totality and apprehended as a geography of experience and movement. By way of conclusion the point can be shown in two ways, one in relation to Ronsard of the *Discours* and the other to the greater themes of space and place. In a discussion of the "variety and vanity of opinions" concerning religion, Montaigne invokes the *Remonstrance au peuple de France,* published late in 1562 (L 11, 63–75; PL 2, 1020–39), in which Ronsard imagined how a pagan, an ardent believer in the sun, would see in this deity a "common light, the eye of the world" (494/515).[18] As cited in the text, the sun of an inalterably *other* religion but eminently familiar in the way it fits in the armillary machinery of cosmography—

> Ce beau, ce grand soleil qui nous faict les saisons,
> Selon qu'il entre ou sort de ses douze maisons (*Remonstrance,* L 11, 65;
>   PL 2, 1022)
>
> [This beauteous, great sun that makes the seasons
> As it enters or exits from its twelve houses]

—is paradoxically "the piece of this machine," so near—familiar to any and every reader of Apian or Münster—but, adds Montaigne, so far. Even though the sun is "full of immense grandeur, round, errant and firm" it is not immediately accessible. Like the Ronsard of the "Discours contre Fortune," Montaigne thinks through a new or another world to bring relativity to religion and credulity. In the greater context of the "Apologie" he is varying on the minuscule but vital difference he discerns between *religion* and *region.* The former, which would be the dominion of the cosmographer, is countered by the latter, the domain of the topographer. The slippage between the terms becomes glaringly clear when, at the beginning, like Apian's or Gemma Frisius's surveyor, he takes a place—any place—to be a site into which one is born or from which a person takes cognizance of his or her position in the world at large. He notes that

> nous ne recevons nostre religion qu'à nostre façon et par nos mains, et non autrement que comme les autres religions se reçoyvent. Nous nous sommes rencontrez au païs où elle estoit en usage; ou nous regardons son ancienneté ou l'authorité

des hommes qui l'ont maintenue; ou creignons les menaces qu'ell'attache aux
mescreans; ou suyvons ses promesses. (422/447)

[we receive our religion only in our fashion and by our own hands, and not
other than as other religions are received. We are met in the country where it
is practiced; or we look at its tradition or at the authority of the men who have
upheld it; or fear the menaces it attaches to miscreants; or follow its promises.]

The litany of choices made available to adduce the fact—our gaze upon
the depth of a foundation or its famous practitioners, our fear of the
threats it poses, our will to believe in its promises—is carried not by a
grammar of *either/or* but of *one/and,* as well as *and/and,* which can run
to infinity. Crucial is the homonymy of *ou* and *où,* of "or" and "where":
every clause that follows the invocation of the pagan world, *le païs,* allows the optative conjunction to be identical to the locational relative
that can be reiterated ad infinitum.

It is here that the cosmographic scope and volume of the essay reveal
its topographical inverse. Montaigne goes on to note that after we engage a religion as we do with our hands, in a sort of ritual meeting that
invokes the handshake of an "other" and oneself *(par nos mains),* the
considerations that follow are "human linkages," which are all subsid-
iary to belief in general. The latter is strong and true but protean. "Une
autre religion, d'autres tesmoings, pareilles promesses et menasses nous
pourroyent imprimer par mesme voye une croyance contraire" (422/447).
[Another religion, other witnesses, similar promises and menace might
impress upon us by the same way a contrary belief.] The beguiling relativity of the phrasing that responds to the elements declined in the preceding sentences is based on a sense of regional difference. Rather than
following the hypothetical thread of the argument set forward in the
text of 1580 (in the "A" stratum of the text), Montaigne adds, in 1588
(in the "B" layer that includes the third book), a famous *aparté* that
refers to things pagan and things regional: "Nous sommes Chrestiens à
mesme titre que nous sommes ou Perigordins ou Alemans" (422/447).
[We are Christians in the same measure as we are either Perigordians or
Germans.] It would not be wrong to see in the geography of the statement not only the copresence of Catholic and reformed views but also,
in the given context, the same amphibology that inflects *ou* with coextensive inflections of difference and location.

They are what constitute the living thread of the "Apologie" to the
degree they are tied to the single most employed bundle of words in

the essay: household, lodge, and clockwork or *horlogerie*. From the out-set Montaigne quibbles with the philosopher Herillus, who invested or "lodged" in knowledge *(science)* "sovereign good," something that the au-thor's father wanted to welcome into his household, and that when the eminent guest Pierre Bunel took leave, *au desloger,* left with the family the book that the author would later translate and for which he would write an impassioned apologia. The coming and going with which the author inaugurates the essay finds its analogue in the way he pictures the swallows under the rooftop of the house. Their nest, like that of the hal-cyon, becomes the secret space that is everywhere the event of the essay itself. The surveyor who is the author is both in and outside the space he has created.

Conclusion: A Tactile Eye

The reader of Monique Pelletier's exhaustive study (2009) of the development of the maps of France and its provinces from early printed editions of Ptolemy at the end of the fifteenth century to François de la Guillotière's meticulously detailed *Charte de France* (drawn from 1570 to 1584, but published over thirty years later) immediately discerns the rapid evolution of topography from a condition of abstraction to that of close and painstaking representation of the French nation and its provinces. In the course of a century depictions of the greater world appear quickly superseded by many more of the places in which their authors found themselves. Following a similar axis, but conveyed in a spirit of speculation, the work in the preceding chapters contends that artists and poets of Renaissance France, who are sometimes themselves draftsmen and geographers, create topographies of sensation and experience. They identify with given regions, they affiliate with new geographies of the nation, and they fashion creations that rival the world in which they are found. Further, and from the standpoint of the ways that we cast our eyes upon their works, we find these writers creating *events* in which the mix and stir of local and cosmic space and time are brought forward. Ultimately, like Montaigne in his *Essais,* they reach into themselves to locate where and how they move in an unstable and conflicted world, in the turmoil of what he famously called *une branloire perenne,* a place whose measure and limit he had shown to be beyond anyone's ken.

The authors and works taken up here have been set under a totem, the snail of the twentieth emblem of Gilles Corrozet's *Hécatomgraphie* (see Figure 1). The common gastropod, often seen on the vignettes and in the marginalia of illuminated manuscripts, emerges from a stony lair as it moves into the landscape in the light of day. Resembling the pupil of an eye, it goes forward while, staring at us, the spiral line of its shell form turns inward, back, and around itself, into what we might imagine to be the *profondeurs opaques de ses replis internes*, the "opaque depths of the inner folds" of itself. The snail whose shell casts its gaze upon us looks ahead as it were, with a sense of touch. Its erect antennae feel the atmosphere about it, and even if we throttle our imaginations, we sense that its proboscis moistens the ground that could indeed be the surface of the paper on which Corrozet's woodcut is printed. The pupil of its spiral shell signals that the very gaze it casts upon us happens to be its home, indeed the world in which it lives. It moves outward tactfully, as Corrozet's gloss reveals, carrying with itself its own "secret," which we can take to be a topography that becomes ours to fathom in the context of the *Hécatomgraphie* and beyond.

Set adjacent to this epigraph is that of the isolated eye (Figures 2 and 3) in the celebrated woodcut with which Pieter Apian illustrates what distinguishes geography from topography. His image makes explicit an analogy inherited from the first sentences of Ptolemy's *Geographia*, which is drawn and explicated in the *Cosmographia* in order, seemingly, to cement in memory the relation of a part to a whole: a world to a face; a city, perched on an island floating on a sea, to an anatomical eye and ear. Yet as we gaze upon the isolated eye we discover that, far from the body above, it is indeed *detached* from the world in which it is found—at the same time that upon closer view the cross-hatchings that depict the organ assume the form of a polar projection of the world, indeed of the kind (Figure 13), seen later in the manual, that includes the European, Asian, African, and American continents. Bereft of the socket in which it would be found, the topographical eye exudes strangeness, an alterity at once to itself or its body and to its milieu. The linkage it convokes between itself and the world at large is shown in the division between hearing and seeing and that of being and moving. What a patient "reading" reveals of the ocular shape of the snail and the whole-and-part of Apian's eye—their partial and local character—pertains directly to the poets and writers of topography.

Both images, it has been argued, draw attention to the tactile virtue of the gaze they emblazon. In the analogy of the world map to the subject of a portrait and of a city view to the detached eye and ear (in both Pieter Apian's *Cosmographia* and Gemma Frisius's additions and emendations), variously emotive—moving, sometimes secret—relations are established among toponyms as they are indexed, noted, and described. The slim but copiously illustrated manual offers ways of thinking that in the Ptolemaic universe can lead to mosaic geographies of subjectivity. Where the cosmos is figured, so also is the bodily presence of the geographer who represents local places. The work reads as a guide for the understanding of inherited notions of cosmography that teeter in the direction of questions concerning what it means to reside or "be lodged" in a given place. The gist of Apian's work suggests that cosmography becomes a function of topography; the latter, which had been subjacent to the former, seems to supersede it in importance and, as a prevailing topic in the manual, opens onto broader questions about the art of observation. Yet the book everywhere displays signs of detachment and dismemberment that correlate bodily isolation with that of topography in respect to cosmography. Hands that are drawn for indexical purpose become as strange as the detached eye and ear that depict the topography. Holding pieces of string to illustrate how a topographical map is drawn, or grasping the handles of globes as if they were consoles, hands are part and parcel of webbings of lines traced to distinguish cosmography and topography. Disembodied wind heads are attached to bowel-like clouds in the margins of Frisius's truncated cordiform map (Figure 15) inserted in editions of the middle 1540s. Trigonometry and triangulation are brought into local landscapes in which the surveyors can be imagined taking stock both of the land and of the positions they occupy when moving about within it. Because of its ties with Martin Waldseemüller and to Amerigo Vespucci's letters, the *Cosmographia* engages a brief and promising description of new and strange places.

We discover in retrospect that Apian's topographical impulse shares much with Rabelais's *Pantagruel,* in which an encounter with alterity, an event in the strong philosophical sense, occurs in a familiar landscape held within an uncommon world. In his exchange with a humble cabbage planter, the narrator lets his inquisitive words and their response mediate whatever fear of difference would be felt at the moment of the discovery of a new world. Alterity is negotiated, indeed "introjected" in an analytical sense, in the form of the account whose analogues are all about and around it, including both the episode of Epistemon's *coupe testée,* in which the teacher, having been decapitated, recounts his ventures in the carnival of hell, and that of the descent of diggers who venture into the hero's bowels to clear them of their fecal blockage. The book moves to and fro between worlds known and unknown, and in its design it appears to draw direct inspiration from Oronce Finé's double (indeed binocular) cordiform projection of 1531, a prescient world map in which a divided world is shown such that one-half of the world, as Rabelais's narrator says, is isolated from the other. In that divided—but also variegated—space, Rabelais eventually inscribes his own signature. A world map becomes the background for topographical narratives that have anthropological latency: the characters of the world "over here" take pleasure of welcoming those others who are at once within and without the greater world of the hero's body.

The same author made clear that he took sensuous delight in cultivating an art that Alcofribas, chronicler and historian, reports abhorring. The rebus—the strange creation anticipating the figure of catachresis, in proximity to the heraldic world of the emblem, pictogram, and hieroglyph—relates Rabelais to a genre that includes images of the world within familiar landscapes. The rebus brings its practitioner into the material world, here and now, of local time and space. In view of Apian and of passing remarks in *Pantagruel* and *Gargantua,* the simultaneous division and synthesis of a figure of speech or a device and an image invite study of the topography of the emblem. The frequently minuscule pictures in emblem books contain landscapes that suggest the presence and the locale of their origin of production. Through a miniature greater worlds are made manifest. In Gilles Corrozet's *Hécatomgraphie* the "world" and "fortune" are correlated, not only in accord with scripture to foster faith, hope, and charity (as in Figure 16) in the sublunar world (as in Figure 20), but also to experiment with various ways of constructing representations of space through combinations of poetry and images. For Corrozet's *Simulachres & histories de la mort,* Hans Holbein draws prototypical emblems in flawless detail and with uncanny wit. Often the art of engraving melds with that of the landscapes engraved. Cutting into the images, the edges of printed letters in the surrounding text turn the book into a topography that is shown to be located in a greater but generally unfathomable world at large. When the mode of production of the image is seen within the space of the engraved picture—when it cuts its furrows, like the celebrated plowman of the *Simulachres* (Figure 26)—the complex relation of the image to the world is made manifest.

Through emblems Maurice Scève at once locates and is lost to his *Délie,* "object of the highest virtue." Comparison of the polar projection of the known world on Pieter Apian's "cosmographical mirror" to the emblem and poem comparing the author to a spider at the core of the webbing of his work attests to a cartographic latency. In the first French *canzoniere* the poem and the image to which it subscribes often become a mental map and a mix of topography and the tensions of mental or psychic mapping. In *Saulsaye,* Scève's eclogue published three years after *Délie,* two woodcut images by Bernard Salomon inflect the poem with uncommon spatial intensity. Of size and aspect resembling emblematic settings or windows, they look onto the world of the poem that situates itself between a contemporary site—Lyons, 1547, at the foot of Mount Fourvière—and classical sources out of which it draws much of its inspiration.

Scève's *genius loci* is not lost on Ronsard, leader of the Pléiade in the middle years of sixteenth-century France, roughly between the heyday of Petrarch and Petrarchism and the beginnings of the Wars of Religion. The *Amours* tie experiments of style and the locative traits of the author's signature to the art of impression and of engraving. As in many emblems, the poems are incisively in and of the landscapes they describe. They might be called fragmentary "event maps" endowed with wit and spatial invention. And they become increasingly geographical in the *Continuation* of 1555 but soon after are mixed and altered in the new forms brought forward in the second book of *Meslanges* of 1559. In that collection, two keystone poems, the one titled *Elegie* and the other *Complainte* (both soon after becoming qualified as *Discours*), chart new territories. The poet takes stock of where he finds himself in political and social turmoil. He writes circumstantial verse, but he also draws attention to the New World and its colonization. A political ethos is born of doubt over what constitutes the limits of the world at large and the places that different people inhabit. The work that precedes the more committed and polemical poetry of 1562 and afterward displays a self, a strongly marked "I" obliged to invent a poetic space in which he can live. As he looks inwardly and to his own world, he also becomes a topographer of the kind Montaigne would soon reclaim in his essay on "Cannibals."

In the opening pages it was remarked that a historiated initial bearing an image of Ronsard inaugurates the third volume of Montaigne's

Essais of 1595. The poet's portrait in the letter that begins "Personne n'est exempt de dire des fadaises: le malheur, est de les dire curieusement" (767/790; see note 5 for the Introduction regarding the split references to the *Essais*) [No one is exempt from uttering banalities: the misfortune is in uttering them with too much care] pictures Ronsard at an easel where he draws from inspiration with one foot on the globe. Montaigne, who took pride less as a poet than for his poetic gait— "J'ayme l'alleure poétique, à sauts et gambades" (976/994) [I love the poetic fashion, in jumps and starts] he wrote in "De la vanité" ["Of Vanity"]—invented an art that he called "legere, volage, demoniacle" (ibid.) [light, flighty, divine] that in prose would be the equal of the bard in his verse. Montaigne creates a *style* and a space in the project of self-portraiture, which, when Apian's images are recalled, make each essay become a fragmentary part—an eye, an ear, a birthmark, a wart on the nose—of the greater verbal *mappa mundi* of himself. Each essay would be likened to a region, and each region would include local images of ways of being, of contending and of living with the world. But the ideal totality of the portrait, like the world that Ronsard's foot happens to touch, is never present or whole. A cosmography of a self-portrait is what the essays are not: they are collections of remote experiences, scattered events from different times and spaces, with which the writing subject seeks to gain a linkage and a relation with a world welcomed for the uncertainties and the creative doubt it inspires.

In the "Apologie de Raimond Sebond," the most ostensibly "cosmographic" essay in the volume, a sense of detachment and of self-isolation is paramount. "Nous n'avons aucune communication à l'estre" (586/601) [We have no relation with being], he famously noted, but in such a way that being is compared to the flow of the world's waters, the very milieu in which he found the fabled halcyon's nest, the home and hearth—like that of the shell of Corrozet's snail—in which a creature can live. The bird's nest bears resemblance to the "lodgings" that swallows construct under the cornices and between the crenellations of his tower. In those birds he sees "local" ways of living and doing. His descriptions of their "being" tip in the direction of new topographies, new senses of ways of living, in worlds from which humans have detached themselves. In the rhapsody of birds and beasts that comprises much of the longest of all the essays are encountered various senses of place, of region, and of an art of living. If the essay embodies a crisis of skepticism it also carries a sensuous and even diabolical pleasure in what it describes of the ambient world. Itself an open whole, what modern philosophers would call a *Tout ouvert* or francophone poet a *Tout monde,* the essay stages a spatial crisis that its own writing tends to resolve in "rolling and flowing" about the flora and fauna of creation.

It has seemed appropriate to end the study with the most aberrant, and also the most errant, and sensuous too, of all of the *Essais.* In the "Apologie" Montaigne rehearses crises of region and religion, and with them he also crafts *other spaces* in which he invites his readers to get lost. With the absence of cardinal direction a poetry of topography is begun. It is shaped further and in different ways in the later essays that can forever be the topic of further and closer study. Yet given the snail's space and pace of the preceding chapters, suffice it to speculate on what event or poetry of topography becomes: more and more a relation of experience felt at once remote and present, a relation in a world that, both today and as it is shown in the gist of the "Apologie," is in decline and diminution. The return to the local worlds and mappings of the poets taken up in this study follows the line of the ocular spiral of the snail's shell: inward and outward, from a place touched and felt, to areas unfathomed, unknown, seen at once here and there, in the confines of poetry acquiring commanding force when brought into our own time, space, and place.

Notes

Introduction

1. Francesco Colonna, *Hypnerotomachie* (1546), fol. 20r. Here and elsewhere all translations are mine. Colonna's allusion to the snail stands in fitting contrast to an image above the colophon in the 1541 (Troyes) edition of *Le grand calendier et compost des bergers.* The diminutive gastropod, perched on a plinth that stands in front of a château, defies two soldiers who wield swords in front of its antennae and a woman who threatens it with a baton. It responds to their menacing words, stating, "De ma maison me suys armé / De mes cornes embastonné . . . je cuyde en bonne foy / Quilz tremblent de grand paour devant moy" (I am armed with my house / And am armed with my horns . . . I truly believe / that before me they shiver with fear). In the almanac the snail is in comic warfare while in Colonna it is endowed with haptic virtue.

2. Building on David Lindberg (1981) and Robert Nelson (2000), Anne Dunlop notes that sight "was essentially a form of touch, with the mind in contact with the persons and spaces around it through the layers of the eye" (2009, 119). Her careful study of the visual experience of mural painting in early Renaissance Italy clearly recalls Alois Riegl's notion of the "haptic" experience of space (1985), which also informs the style of readings to follow. Roland Greene (1999, 80), using Michel de Certeau (1975), takes up the colonial inflection of sight, according to which what is found in the field of view is objectified at the very moment the visuality or "point of view" is made clear.

3. A grounding principle of Michel de Certeau's work on the history of mystical experience (1982) is that when it is named as such (in the seventeenth century), it becomes a scientific object. In its adjectival form the existential character is retained, but when it is under the banner of the substantive "mysticism," as a scientific object

its force of attraction is nearly dead and gone. Certeau's words share affinities with a vital relation with the unknown that Rosolato (1995, 153–69) studies extensively through what can be known *(un inconnu connaissable)* and what cannot *(un inconnu inconnaissable).*

4. See Gilles Deleuze, "What Is an Event?" in *Le pli,* 100–103.

5. Readings of the chapter are legion. Some are enumerated in my *Graphic Unconscious in Early Modern Writing,* 116–34, from which I would like to diverge here.

6. Michel de Montaigne (357/377). References in text are to the 1962 edition of the *Oeuvres complètes* (Rat and Thibaudet) and the 1988 edition of the *Essais* (Villey and Saulnier), respectively.

7. See the color facsimile of the "Exemplaire de Bordeaux," edited by Philippe Desan (2002), which shows exactly how much Montaigne worked over the passage.

8. Gilles Deleuze (1992, 72).

9. Deleuze (1988, 106).

10. Evelyn Edson (2007) studies the continuity of the world as a closed totality in medieval cartography. She notes that "Ptolemy conceived of the world as a single and continuous entity" and that "atlases of maps made according to Ptolemy's instructions were," in the later Middle Ages and the Renaissance, "supplemented with new local and world maps," which "accepted, rejected, and revised the originals but invariably followed his model of unornamented, measured space" (234).

11. The subtitle of Corrozet's book aptly describes its contents: *Hécatomgraphie: C'est à dire les descriptions de cent figures & hystoires, contenans plusieurs appothegmes, proverbes, sentences & dictz tant des anciens que des modernes . . .* The printer, Denys Janot, was known for his consummately illustrated books, especially those of emblems.

12. Frank Lestringant (2006, 263–66) studies the anthropomorphic landscape, which pertains to the image of the snail through Pierre Belon's *Observations de plusieurs singularitez et choses memorables . . .* (Paris: Gilles Corrizet, 1553).

13. It is noteworthy that the form of the shrub resembling the snail is also found in the anthropomorphic design of the third illustration (fol. 4v) of *Le songe de Poliphile.* The artist drawing the image in the work of 1546 no doubt uses the landscape of this emblem to fashion the background of the remarkable picture of Poliphile in tears at the base of a tortuous oak tree. Furthermore, the spiral of the shell as Corrozet displays it could be related to the logarithmic spirals that pertain to navigational loxodromes in, taken up in the chapter on winds in Pieter Apian's *Cosmographia* (in Hallyn 2008, 98–100). The relation requires further study.

14. T-O maps were iconic schemata by which the continents, the sons of Noah, who inherited them, and the major bodies of the world's waters were distinguished by a *T* inserted in the lower half of an *O.*

15. That maps are objects of contemplation is a richly discussed topic. Giorgio Mangani (2006) offers a history of the energies invested in the meditation on maps; Kathy Lavezzo (2006) tilts meditation in the direction of ideology—conquest and nationhood. Gunnar Ollson (2007) reads the Ebstorf map as an object that inspires visual meditation. Michel de Certeau (1982) determined that prior to oceanic travel the map could prompt quasi-mystical voyages by the way the viewer saw himself or herself traveling about the world while "reading" the map itself. Francesca Fiorani (2005, 99) notes in her reading of Apian how a separation of one world from another

alters Ptolemy's model of a terrestrial world enclosed in a celestial matrix. Elsewhere I have tied the image to a project of the experience of meditation (Conley 2009).

16. Meyer Schapiro (1973) notes that the Pantocrator who looks directly at the viewer in medieval images gains power through a gaze that humbles whoever is in its field of view ("the Pantocrator needs you"). By contrast, images of people seen praying in profile confer power upon the viewer whose gaze is cast upon them without the disturbance of visual acknowledgment or response.

17. The image of Christ in profile is found in Italian medals in the fifteenth century. Wood (2008) notes that the profile format evinced "pure contour" and was "apprehensive, publishable"; it was form as a kind of writing that "carried strong connotations of authenticity, proposing an authoritative set of features" (155–56).

18. In his rich and richly documented *The English Renaissance Stage,* Henry Turner (2006) brings geometry, cartography, and urban planning into the perspective of theater. He coins the term *topographesis* to apply to plays "in which the representation of location becomes the primary mode of giving structure and sequence to the symbolic content of the performance," adding that "it [discards] the specific way in which any given text integrates the representation of place into the wide variety of interpretive conventions that were typical of different forms, genres, subject matters, or stylistic modes during this period" (31). The definition is built upon the early modern definition of *topographe* and *topographie* that Randle Cotgrave registers as "a describer of places" and "the description of a place," respectively, in his 1611 *Dictionarie of the French and English Tongues.*

19. *Pantagruel,* in Rabelais (1994, 244–45). Subsequent reference to Rabelais will be made to this edition and cited in parentheses in the text.

20. An incomplete and partial reading of this kind is indicated in my *Self-Made Map,* 145–48.

21. The dedicatory epistle, in Latin and in Huchon's modern French translation, is in Rabelais (1994, 988–90). In North America a copy is found in the Houghton Library (FC 5 R1125 534m).

22. For a history and description of the latter see Mireille Pastoureau (1984, 131–33).

23. Antoine de Bertrand (1587).

24. Cécile Alduy (2007, 319–25) alertly remarks how the cordiform frontispiece of Petrarch and Laura drawn for *Il Petrarca* (in the 1547 edition by Jean de Tournes and 1551 edition by Guillaume Rouille) is a model for the 1552 edition that features Ronsard and Cassandra.

25. "Chifres de lettres entrelacees," in Geoffroy Tory (1529, fol. lxxix r).

26. It is worth recalling the letter of 1570 to Michel de L'Hôpital, chancellor of France and accomplished neo-Latin poet, in which Montaigne dedicates Étienne de la Boëtie's Latin verse to the chancellor, and whom he describes as any of a number of geniuses who seek "nothing more *curiously* than the way by which you can arrive at knowing the men under your charge" (1962, 1363). The letter itself is a masterpiece of ornate, delicate, and "curious" rhetoric.

27. Marc Augé (1992); Gilles Deleuze (1983, 286), for whom the original is *espaces quelconques.*

28. Elisabeth Hodges (2008) takes up the city view as a guide for the reading of this essay, especially the rhapsodic pages on Paris (950/958) and Rome (975/983) that are

shown related to the concurrent "book of the cities of the world" (cf. Georg Braun and his *Civitates orbis terrarum;* in French as *Théâtre des cités du monde*).

1. Rabelais

1. The illustrations are taken from the Houghton Library copy (Typ 515.12.463). See also entry 326 in Mortimer (1964, 421–22).

2. The Ebner Codex (circa 1460) by Nicolaus Germanicus (in the New York Public Library) is translated and reproduced in facsimile in Ptolemy (1460/1991). The edition contains a facsimile of a variant of a similar map found in Tory's *Itinerarium* (163; also in Shirley 1983, 8, entry 8). It is noteworthy that the water masses of the world are not shaded with hatching and that no vertical column is seen separating one half of the world from the other.

3. In Rodney Shirley (1987, 18, entry 25, plate 19). Schedel's winds, however, are made to be seen not all about the map but, because they are all right side up, show that the map is to be seen from the standpoint of the three sons of Noah, located in the spandrels, who hold the map and gaze upon it. A variant of the Tory map, dating to 1482, serves as a comparative measure (Shirley, 7, entry 8). The latter appeared in a Venetian edition of Pomponius Mela's *Cosmographia*.

4. Delano-Smith (2006, 35–40) insists on the topographic character of the itinerarium and, along the way, considers the nature of experience that an individual traveler—much like that of Alcofribas in Pantagruel's mouth—might have known in the sixteenth century (24–29).

5. A sense of gridding is key, which brings forward the relation of Ptolemy to the rebirth of classical perspective, for which Samuel Y. Edgerton argues (1987, especially 13–14).

6. See especially Frame (1977) and Stephens (1989) on the plasticity of proportion in the novel; Alfred Glauser (1966) notes that "the most living places in Rabelais are rarely described: they are suggested; they live through the people who inhabit them" (93). Baraz (1983) observes that the gigantic proportions of the work free its reader from "the prosaic and conventional image of the world; now free, the imagination modifies the proportions of things according to its inner necessity, independently of the law of what we ordinarily believe reality to be" (23).

7. Drawing on Geoffroy Tory's *Champ fleury,* Hampton (2001) invokes the figure of the "garden of letters" in which print culture is taken to be a "flowered field" of typographic forms. The setting he describes belongs both to Rabelais and the map the same author includes in his itinerarium.

8. Gunnar Ollson (2007, 163–81) situates Auerbach's *Mimesis* (1953) in a history of displacement and draws from its style and structures an impulse to "map" the ways in which uncertainty determines much of the human condition. Ollson's reading of the Ebstorf map (61–75) can pertain to proportion in Rabelais.

9. The details are illustrated in Shirley (1987, 18, entry 25, plate 19).

10. For Nicolas Abraham and Marie Torok (1978), introjection is associated with the act of internalizing a human relation, of "installing into oneself an object that serves as a point of reference for taking note of *[pour l'appréhension de]* an outer object," and

is based on the fact that "we have the innate faculty of being the subject and object for ourselves. I touch the palace of my mouth with my tongue, I hear the sound I am emitting, I see the movement of my hands that I am causing to move: in general, I am two in one" (127–28). Laplanche and Pontalis (1976, 209–10) define the term with a spatial inflection that bears, as do Abraham's words, on what is called an "event" in my Introduction. Through the labors of his or her fantasies, a person moves from the "outside" to the "inside" of objects and the qualities inhering in these objects. Laplanche and Pontalis recall how (for Sandor Ferenczi) the term was related to transference and how (for Freud) introjection is opposed to projection. A person "introjects" a source of pleasure while projecting outside his or her bodily space—the area where indeed geography begins—what causes displeasure. In "Mourning and Melancholia" (1957, 14:248–49), Freud suggests that incorporation amounts to cannibalization of the other. Inferred is that introjection would be at a remove from incorporation.

11. François de Dainville (1964/2002, 210); Catherine Delano-Smith (2007, 574), especially figures 21.43–44.

12. Such as an analogy of Rabelais's master Galen: in *De usu partium* the cofounder of Greek medicine had described the mouth as a cavity striated with roads, the esophagus as an isthmus, and the stomach as a barn. The whole body was shown to be comparable "to a city of trade in which everyone minds to his or her business," notes Robert Antonioli, in *Rabelais et la médecine* (155; cited in Rabelais 1994, 1336n7).

13. Huchon sees a variant of an episode related in Masucio Guardati's *Il novelino* (1476) and a salacious satire of men's adieus to women of lighter ways, a genre with which Clément Marot was associated (Rabelais 1994, 1314–15n7).

14. Illustrated, respectively, in Shirley (1987, 74–75, entry 67, and 72–73, entry 66). The innovations of Finé's double cordiform projection are taken up in Frank Lestringant and Monique Pelletier (2007, 1466); the latent presence of the Copernican system, shown by two angels cranking their windlasses at the north and south poles of Münster's map, is studied in Denis Cosgrove (2007, 69). (According to Pelletier 2009, Oronce Fine is the proper spelling. However, based on Karrow 1993, and for consistency with *The Self-Made Map,* I have stayed with Finé.)

15. Tory glosses the perspectival dimension of the letter in the second book of his *Champ fleury* (1529, fol. D.iiii v), a work that indirectly explains much of the graphic tensions in the early Rabelais. It is Maurice Blanchot who draws the visual and vocal virtues out of *trouver* in "Parler, ce n'est pas voir," a dialogue in *L'entretien infini* (1969, 34–36).

16. In arguing for Montaigne's uncommon awareness of the New World, Claude Lévi-Strauss (1991, 277–78) notes how Lucien Febvre (in his groundbreaking *Problème de l'incroyance au XVIe siècle* of 1943) underscores the blindness of cosmographers in view of the impact of the Columbian encounters no less than four decades after Vespucci's voyages.

17. The question concerning the limits of Christendom pervades the geographical knowledge that Rabelais inherits. In Amerigo Vespucci's letter recounting his first voyage, the traveler notes that classical authors do not mention any continental mass or presence of islands west of the point of his departure from Cádiz, which even "Dante, our poet, in the twenty-sixth chapter of his *Inferno,* where the tale of Ulysses's death is found, is of the same opinion" (Ronsin 1991, 85). Lester (2009, 102–9) develops

the medieval perspective on the limits, while Wey Gómez (2008, 115–16) shows how a pre-Christian, Homeric worldview of lands unknown is adapted to the Christian imagination that shapes Christopher Columbus. Duval (1991) establishes a religious design in Pantagruel through much of what pervades these issues of theology and geography.

18. Migration is not mentioned in the text, nor are pigeons noted as carriers of messages. Yet the sense of a greater and broader world is felt in the idea of migration: birds go "there," to places one knows not where; yet in their absence they are present, "here," in an invisible or even mystical way. The experience of migration is taken up in chapter 6.

19. At the very heart of Oronce Finé's double cordiform world map in Grynaeus's *Novum orbis regionum* (1532), the book Rabelais is said to have consulted, Cathay is located beyond Florida and the Gulf Coast, attached to the Asian continent. In view of this map Rabelais's geography seems quite clear.

20. Among others, Francesco Rosselli's small world map of 1508 (in Woodward 2007, 14, plate 16) is decisive for the reason that it locates the Hispanic islands on the flat plane of the terrestrial sphere.

21. The formula *puis* marks the voyages of the younger Pantagruel, in chapter 5, when he travels from one city to the next. Early in the work the term connects the places visited, in "L'enfance de Pantagruel" (chapter 4), in what Frank Lestringant called a "toponymical tale" (1993, 111).

22. Pigafetta's account and its complex history are taken up in Cachey (2007a). It clearly informs Finé's maps and Rabelais's geography.

23. After Pantagruel cuts the Gordian knot of the cock-and-bull debate between Humevesne and Baisecul, "le jugement de Pantragruel fut *incontinent* sceu et entendu de *tout le monde*" (262) [Pantagruel's judgment was suddenly known and understood by the whole world].

24. Chapter 17 of Apian's *Cosmographiae introductio* (1535) is titled "Quo differunt, insula, peninsula, isthmus & continens" (fol. 28r). The distinction among the four major bodies seems to ground the play on proportion here and elsewhere.

25. In this manner the play on words confirms Terence Cave's hypotheses concerning the void over which *copia,* indeed, a plethora of information, is written; see *The Cornucopian Text* (1979).

26. One way of seeing *tout* is through the scheme of the T-O map that had been an iconic *mappa mundi* in the High Middle Ages. The T and O of the *Orbis theatrum* conjoined orthogonal and circular lines. *T* was a trumeau and a lintel within an *O,* a *tout,* that divided the world into its three continents, showed that they were apportioned to the three sons of Noah and were situated within the surround of the *ecoumene,* or circular border, of a *mare oceanum.* The first printed T-O map is found in an edition of Isidore of Seville's *Etymologiae* (Augsburg, 1473). The iconography of its composite letters are taken up in Zumthor (1993) and are studied in their spatial latency in Tory (1529, fol. xxii, li–lii, and lxx) without specific reference to the *mappa mundi.*

27. The model is found in other pairings and doubling of chapters, especially vi–ix, which enframe vii–viii (Conley 1996, 153).

28. See Rabelais (1994, 328n4).

29. It is common in infernal diagrams, such as what Ricardo Padrón (2007) takes up.

30. Lemaire's poem was written to build on the celebrity of the "Première epistre" and printed in the first (and ongoing) edition of the poet's collected works (1512–13). In this edition the first and second poems are inserted between the end of the first and the beginning of the second volumes of the *Illustrations*. Their presence on the title page and placement in the volume attest to their renown.

31. In *Le Moyen de parvenir* (Béroalde de Verville 1612?/2006, 174–75), an anecdote that mocks the judgment of Paris tells of a man who asks three young women to explain why the mouth is either older or younger than the sex. The prettiest and ablest of the three notes that her mouth is older because it is weaned *(sevrée)* and her sex suckles *(mon con teste)* with great pleasure. The fable builds on the wit of the play on words.

32. And the care with which the author edited and altered the compendium from one edition to the next indicates that it was read closely and with an eye to design that changed according to the politics of the moment (Rabelais 1994, 1328n2). Six names are added in 1533; in 1534 the mocking wording about Epictetus and the archer Baignolet is appended, and twenty names are inserted to generate explosive poetry at the beginning, along with twelve others; four are exchanged for others; Roland and Oliver of the *Song of Roland* appear in 1535. In 1542 four more exchanges of roles among the other peers of France are added. Huchon notes a strong political motivation in the editions of 1534: now absent, the kings and peers of France appear immune to the effects of satire.

33. In a general discussion of these chapters, Emmanuel Naya (2008, 175–92, especially 185–86) notes how Rabelais responds to a "loss of points of reference that had organized the space of meaning" (175) by infusing thought with laughter; by staging uncanniness in familiar situations; finally, by reviving skeptical inquiry that "de-centers" the production and interpretation of discourse. He sees that in these pages, *Pantagruel* organizes the concurrent construction and deconstruction of its hero as a model of sanctity. What is taken to bring perplexity to otherwise typically moralizing scenes of chivalric novels can be owed, it is argued here, to what new geography does to inherited doctrine.

34. They belong to a Gothic world of writing, including the *lettre bâtarde,* foreign to the new-old style of the classical character. Here and elsewhere the book seems to belong to what it discovers to be two contemporaneous modes of reading and writing, hence of two different cultures that constitute a new space. The "Gothic" and the "antique" are mixed, just as different modes of typography are found on the same page (Martin 1999, 221–24; Zerner 1996, 14–16).

2. The Apian Way

1. *La cosmographie de Pierre Apian* (1553, fol. 19r). This edition and its Latin counterpart, also published in Paris in 1551 and 1553, will be the major point of reference for this chapter. Comparative references will be made to the first French edition (Antwerp, 1544).

2. This illustration is vital for what Rabelais takes up in *Pantagruel* about the differences between French leagues and their German, Italian, and Spanish counterparts.

Panurge's famous anecdote in chapter 23 about the discrepancy between the bodily desires of the men and women who set out to survey the roads of the nation has its visual analogue in this very book.

3. Certeau (1990, 168–70). The absent body or parts of a body that generate a *genius loci* are taken up in his work on space and hagiography in which he observes that a "saint's life is a composition of places" filled with signs indicating the presence of an absence (Certeau 1975, 285–87).

4. "What in fact makes a map a map seems to be its quality of representing a locality," Buisseret notes, adding, "perhaps indeed we should call it a 'locational image,' or even a 'locational surrogate'" (2003, xi). The locational image stands in contrast to a landscape painting in that its function is to convey information pertaining to the place shown, where its counterpart "primarily seeks aesthetic effect" (ibid.). The contrast is further underscored by way of Ptolemy. Buisseret notes that the Alexandrian geography, drawing on Marinos of Tyre, "wanted to insist that his were directions for drawing a map of the known world, and not local maps. As he put it, local cartography depends on landscape drawing and needs no mathematical method" (16).

5. Cosgrove (2001, 111–12) notes Waldseemüller's influence. A facsimile of the latter's work is found in Ronsin (1991). Elsewhere Cosgrove (2007, 67) notes that the *Cosmographia* "was more popular than original and fuller than Waldseemüller's" insofar as it contained "instruction on celestial observation and the practicalities of spatial survey and delineation intended to clarify the accompanying world map" in the editions Gemma Frisius edited after 1530. Frisius's life and works are studied with care in Hallyn (2008).

6. Karrow (1993, 49–63) offers the most cogent and readily available biography. Apian (1495–1552) was born in Leisnig (between Leipzig and Dresden) and studied at the University of Leipzig before moving to Vienna in 1520. His excellence in mathematics and astronomy was evinced in his first world map of the same year, based on Ptolemy and Waldseemüller (in Shirley 1983, 51, entry 45). His *Isagoge* (1521) explains the map and anticipates his *Cosmographicus liber* (Landshut, 1524) that would go through twenty-nine more editions over the next eighty-five years. In 1526 Apian and his brother Georg settled in Ingolstadt, where they opened a print shop. The following year Pieter was awarded a professorship of mathematics at the university. He published numerous maps and manuals, among others the *Cosmographiae introductio* of 1529 and 1532. His cordiform map of 1530 preceded two books on the quadrant and horoscope. He discovered five comets between 1531 and 1539, prior to publishing his *Astronomicum Caesareum* of 1540, ostensibly his greatest work in the science of the heavens, "one of the last great attempts at explaining the Ptolemaic system" (Karrow 1993, 62). It is well known that Apian found a friend in Charles V, the emperor of Spain, who named him the court mathematician. Charles then dubbed him a knight of the Holy Roman Empire. As Karrow notes, Apian's "real contribution to the history of cartography lay in his *immensely popular books* on geography, cosmography, and astronomy and in his development of observational instruments. He was also a printer of extraordinary books, an aspect of his career that has not been sufficiently studied" (62; emphasis added).

7. See Finé's "La composition et usaige d'un singulier métyhéroscope géographique" (1543), illustrated in Frank Lestringant and Monique Pelletier (2007, 1465, Fig. 47.1);

see also Finé (1542, 1549, 1551); Focard (1546), Foullon (1561), and Merliers (1575). Uta Lindgren (2007) offers a rich review. She notes that in his treatise appended to the *Cosmographia,* Gemma Frisius brings forward a first practical treatment of triangulation after Alberti and Regiomontanus. By and large, however, a "lack of correspondence between theory and practice in land surveying mirrors a similar lag in the general cartography of the Renaissance, where modern methods of compiling maps had been postulated long before observations of sufficient precision were possible" (508). Imprecision is implied to belong to the aesthetic regime of the landscape, to what the painter and writer take as their task to describe.

8. The summary (emphasized in text) reflects the title so as to make a closed whole of the book itself. See Figure 10. The editions consulted here belong to the Bibliothèque Ceccano, Avignon, shelf number: 8 15624; and the James Ford Bell Library at the University of Minnesota, shelf number: TC Wilson Library Bell 1551 Ap. The earlier French edition—*La cosmographie de Pierre Apian, libvre tresutile . . .* (1544; Houghton Library, Harvard University, *GC5 Ap34 Eh544b)—on which Gaulterot's new version is based, omits reference to the last clause "the whole with appropriate figures in order to provide intelligence." The Parisian editions, it will be seen, amplify and stress further the "locatedness" of the *Cosmographia.*

9. "Si le monde voullez scavoir pourtraire / Et le circuit de la terre perlustrer /Che *[sic]* livre achaptez, il le vous declaire / Si le monde voullez savoir pourtraire / Sans aller loing, & grand despend faire . . ." (fol. ai v) [If you want to know how to portray the world / And travel about and around it / Buy this book, it will tell you. / If you want to know how to portray the world / Without going far and at great expense . . .].

10. E.g., *Cosmographia Petri Apiani . . .* (Antwerp: Gregorio Bontio, 1550), fol. ai v, and (Paris: Vivantium Gaultherot, 1553), fol. ai v.

11. The Latin of the Paris edition does not use this typographic design in its layout of the prefatory letter to the first edition (1524), but the 1550 Antwerp edition does. In both editions the names of the dedicators and the dedicatee are set in the cul-de-lampe format.

12. Editions from Antwerp (see nn. 6 and 8 for this chapter) illustrate the title page with an image of a mounted globe that figures at the end of chapter 14, "Comment on pourra appliquer le globe cosmographique aux quatre coings du monde & tellement qu'il puisse servir à toutes Regions, Provinces, & Villes" (fol. 22r–v); and, "Quomodo globus cosmograph. Ad mundi cardines, & ad quancunq; Regionem, Provinciam, aut Oppidum, rectè sit aptandus" (fol. 20r–21v) [How the cosmographic globe can be applied to the four corners of the world and such that it can serve all regions, provinces and cities]. On the title page the legs of the globe's armature are set firmly on the ground on which a compass and plumb are set adjacent to a cylinder destined to measure latitudes. The globe displays the three known continents on which its names are placed (along with notation of Taprobana) in a sunlit area. The animals that decorate the armature yield an effect of wonder and enigma to what in the text becomes a scientific object where cosmography and topography are met. The style of the globe is to be compared with Holbein's emblem of the cosmographer in his studio in *Les simulachres & histories faces de la mort* (see Figure 23, this volume).

13. The Latin text is as follows: "Geographia (ut Vernerus in Paraphrasi ait) est telluris ipsius praecipuarum ac cognitarum partium, quatenus, ex eis totus cognitúsque

terrarum constituitur: & insigniorum quorumlibet, quae huiusmodi telluris partibus cohaerent, formula quadam ac picturae imitatio" (Paris edition, 1553, fol. iv–2r).

14. In a pathfinding comparison of eyes in Romanesque and Gothic sculpture, André Malraux (1951, 35) notes that where the former, almost exophthalmic, look outward, as if aghast at a cataclysm to come, the latter squint the better to locate and measure the proportion of things in his midst. Further, Meyer Schapiro observes (1973, 1996) that a relation of power can be implied where the figure in profile cannot return his or her gaze upon the viewer.

15. While citing the Paris edition of 1553, I have set between brackets the French of the 1544 edition of Antwerp. In Latin: "Omnia siquidem ac ferè minima in eis contenta tradit & prosequitur. Velut portus, villas, populos, rivulorum quoque decursus, & quaecunque alia illis finitima, ut sunt aedificia, domus, turres, moenia, &c" (fol. 2r).

16. In Latin: "Finis verò eiusdem in effigienda partilius loci similitudine consummabitur: veluti si pictor alquis aurem tantùm aut oculum designaret depingerétque" (fol. 2r).

17. On the map Paris is close to the 49th parallel and the 24th meridian but listed at 47.55 and 17.8, respectively. Rouen is shown close to the 50th parallel and 21st meridian but noted at 49.0 and 15.5. The discrepancies enhance the aesthetic nature of the topographical project.

18. In the *Elementale cosmographicum* of the *Cosmographiae introductio* (1532), the four fingers and thumb are deployed as a memory image to recall the five zones of the world.

19. A reader of literature cannot fail to reflect on the title of Antonin Artaud's early and wondrous collection of poems *L'ombilic des limbes* (1925), in which the author portrays the anguish of being alive, which in the terms of Apian's book would be between a polar scar of originary separation *into* the world and the *limbus in extremo immobilis,* that is, the time of life allowed for such reflection.

20. Apian's map of 1520 (285 x 410 mm), glossed and illustrated in Shirley (1983, 51–52, entry 45), is a truncated cordiform projection under the title *Typus Orbis Universalis . . .* Like Waldseemüller's of 1507 (1,320 x 2,360 mm), a miniature of the Alexandrian's "older" world map stands between the busts of Ptolemy and Vespucci and over the great projection (Shirley 1983, 30–31, entry 26). In the spandrels the winds are formed from patterns of scalloped folds out of which the wind heads emerge. Apian's map uses some of the motifs that Gemma Frisius uses in his map of 1543, also illustrated and glossed in Shirley (93, entry 82).

3. A Landscape of Emblems

1. As in Tory's *Champ fleury* (1529, fol. xlii r–v) and in Germain Harouyn, *Book of Hours for the Use of Rome,* fol. N (illustrated in Bagnoli 2009, 81).

2. Rabelais is in fact close in spirit to the sense of the cinematic ideogram that Sergei Eisenstein (1959, 28–44) develops in a dialectical reading of the word and image. The similarity attests to the cinematic latency of Rabelais's writing.

3. Horapollo, *De la signification des notes hieroglyphiques des Aegyptiens* (1543).

4. *La fleur des antiquitez* is published throughout the century (1586). The relation

of the text to its accompanying maps of the work is studied in Hodges (2008, 49–71, esp. 53).

5. See chapter 6, Figure 38. The importance of Belon's *L'histoire* in the history of the illustrated book (and the resemblance that its colored editions share with maps ordered on command) is taken up in Chatelain and Pinon (2000, 259–61).

6. *Hécatomgraphie* (1543/1997). Line 28 of the edition of 1541 lays less stress on ocularity: "Plus richement, comme on faict par raison" emphasizes the aura of the emblem (fol. Aiii r).

7. The reading is of the kind Christian Metz proposes on enunciation and place in his work on cinema in which, in what he calls a "deictic mechanism," the absent or unfixed place of the voice of the words causes the eye to wander and wonder about who speaks and from what position (1991, 17).

8. It is worth recalling that the first chapter of the French edition of Colonna's *Songe de Poliphile* contrasts deciduous and palmate trees (1546, 3v–4r) and that the allegory of a "dead" versus a "leafy" branch of letters crowns Tory's conceit about learning versus ignorance at the end of book 2 of the *Champ fleury* (1529, fol. xxviii r).

9. Henkel and Schöne (1967, 45) include the emblem for the sake of contrast with other representations of the terrestrial sphere in emblem books of other provenance and style. The catalogue is an indispensable work of reference and comparison.

10. In the 1530s and 1540s, the wandering saint was shown bringing a credo or even a viaticum to the allegorical voyager whose life was imagined as a physical and spiritual itinerary. The importance of Pauline iconography in the age of Francis I (1515–47) is taken up in Lecoq (1987). Its importance for material of cartographic substance is studied in Skelton (1966, ix), where the Psalms, like Corinthians, belong to a reformed geography.

11. Among the avatars count the *Quadrins* for which Bernard Salomon, whose work figures in chapter 4, drew illustrations that merit extended comparison.

12. Reference is made to pagination of the facsimile edition (Gundersheimer 1971) followed by the folio in the Harvard Houghton Library copy (Typ 515.38.456).

13. Dekker (2007, 135–36 and 141–47) reviews the history of cosmographic globes and their armatures, including the terrestrial globe seen in Holbein's *Ambassadors* of 1533. The context informs the image of the astrologer.

14. It would not be a great critical leap to see the presence not only of the trick of anamorphosis in the *Ambassadors,* but especially of his prescient illustration of the heliocentric galaxy prior to Copernicus's *De revolutionibus* of 1543. In his decoration of the world map of 1532 in Grynaeus (taken up here in chapter 1), a cherubic *putto* turns the globe on its north-south axis with the aid of a handle attached to a set of cogwheels. In this image two systems seem copresent: that of Ptolemy, as seen in the armillary sphere; and that of Copernicus, but of Copernicus through a play of the arrangement of objects in the image, through their propensity to convey and to resemble letters and words, and in the play of light and darkness.

15. The bright light seen in the window in woodcut 25 has an immediate contrast in the gridding of the window of the miser's bureau in woodcut 26, which is immediately adjacent. His left arm is raised as if both to point at the light in the occluded window and to exclaim surprise before the skeleton who is helping himself to his smorgasbord of coins.

16. It can be said that Corrozet's translation of Job 28 underscores this reading by having *amphibologie,* a term referring to the materiality of language that rhymes with *astrologie:* "Tu dis par Amphibologie / Ce qu'aux autres doibt advenir / Dy moy donc par Astrologie / Quand debvras a moy venir." (You tell by Amphibology / What of others will become? / Tell me by Astrology / When you to me will come.)

17. For John Ruskin, notes Michael Gaudio (2008, 140), "the value of an engraving lies in what it tells about its own making." He names the woodcut "The Last Furrow" and leaves to George Woodberry, in his *History of Wood Engraving,* his impression of the work: "the demonstration of the engraver's labor, a demonstration seen not only in Holbein's actual lines but in his subject matter, plowing, 'the purest type' of engraving" (141–42).

18. The hourglass is not in Creation (1), Original Sin in the Garden of Eden (2), the Expulsion from Eden (3), Bones of all Men (5), the Pope (6), the Empress (10), the Astrologer (27), the Merchant (29) (although a sun clock is in its place), the Seaman (30), the Count (32), the Duchess (36), the Itinerant Peddler (37), or the Last Judgment (40). Its absence in thirteen of the forty-one woodcuts causes it to become an object that the eye seeks but does not always find. It is both a hermeneutic and a heuristic shape that invites the observer to look all about and around the world in search of its presence.

19. The joke was not lost on George Sand. The great writer of the Romantic era launched a program, contra Balzac's realism, that would idealize the world, and in particular the Berrichon region of her origins. In the preface to *La mare au diable* (1846, 9–11), she compares the forlorn and ugly fate of the plowman to the sad souls Balzac depicts in his trenchant realism. She compares the effects of his work to Holbein's picture. "Set beneath a composition by Holbein, this quatrain in old French is of a profound sadness in its naïveté. The woodcut *[gravure]* represents a plowman guiding his plow in the middle of a field. A vast countryside extends in the distance; some sorry cabins are visible; the sun is setting behind a hill. It's the end of a brutal day of labor. The peasant is old, stout, covered with tatters. The four harnessed horses that he follows are thin and tired. The point of the plow sinks into rocky soil that resists." The vanishing point that goes without mention, the site where the real and ideal are said to meet, is marked by grains of *sand* at the passage between the upper and lower spheres. Would it be the turning point that allows her to conclude, "Thus instead of pitiful and frightful Death, marching in his furrow with a whip in his hand, the painter of allegories might place by his side a radiant angel who with full hands sows the blessed grains of wheat upon the manured furrow" (29–30). Would Sand have seen the sign of her nom de plume at the vanishing point of the image that fascinates her? In the spirit of Holbein it is tempting to think so (see also Conley 1996b).

20. The *A* as an emblem of the compass, an illustration of which is found in Tory (1529, fol. G iiii v), is taken up in the study of Maurice Scève (*Délie,* 77) in the following chapter.

21. See Robert Brun (1930, 41 and 54) on the way a northern (Finé) and southern (Tory) influence dominated the illustrated book in France in the first half of the sixteenth century. Butsch (1878/1969) displays their historiated initials.

22. The celebrated image in *La grant dance macabre* has more recently been set on the

cover of *Critical Inquiry* (31.1), "The Arts of Transmission." Citing the colophon, Péricaud (1851, entry 386 for French books) notes that the book was issued February 28, 1499.

4. A Poet in Relief

1. In *The Sovereign Map* (2006, 354–60), Christian Jacob develops the concept of the mental map that may be viable for a reading of Scève. Jacob notes that the mental map "belongs to the individual" and is "of that person's domain and is completely different from those of his or her contemporaries," noting how each individual lives the map so as to have it represent what he or she thinks the world might be. It stands in some contrast to the totalizing gesture in Fredric Jameson's notion of cognitive mapping (1991, 409–10; 1992, 10), although it will be shown that the poet seeks to offer glimpses of totalities in the reflective fragments that are the poems themselves. In the context of cinematic images, memory and topography (Conley 2007, 18–19), I have tried to construct a model of mental mapping. The poet Adrienne Rich may have put it best where she is said to have remarked that our sentient bodies are the origin of any and all geography.

2. Kennedy's assessment of the "site" of Petrarch in Scève and the poets he influenced (2003, 127–30), definitive and enduring, underlines how *Délie* "psychologizes Petrarch's experience" by "unbinding its energies in a prodigious display of verbal inventiveness" (127). Giordano (2010) studies exhaustively the relation of writing to meditation, which is taken to be reflection on where the poet is in an uncertain world.

3. The spiritual automaton belongs to the idiolect of Deleuze (1985, 161–75); it bears on the politics of cinema much as it might apply to the industry of affect and of what can be made of it as it is generated in this poem.

4. Gérard Defaux (2004: 1, clxxxv). Quotations from *Délie* will be drawn from this edition and cited by the letter *D* and the number of the poem (in arabic numerals), followed by its line or lines. Reference to Defaux's gloss will be indicated by the date (2004) and number of the volume.

5. Ibid., clxxvii.

6. "Voi ch'ascoltate in rime sparse il suono / de quei sospiri ond'io nudriva 'l core / in sul mio primo giovenile errore / quan'era in parte altr'uom da quell ch'i'sono . . .": the first lines of the *Canzoniere* are clearly reflected in those of *Délie*.

7. The "error" of inversion is found in Defaux (1:11). It is so pliable that the printer's or editor's mistake can become a graceful invention: when the poet falls in love (and when the editor falls in love with the poem), the world is turned topsy-turvy, and in fact the inversion suggests that the circle of the enclosing frame engages a rotative or twisted reading. The lexical trajectory would be in productive variance—*in errore*—with the Petrarchan inflection of the motto "Entre toutes une parfaite" (Among [females] one perfect), fashioned from the *Canzoniere,* "Che fu sola a 'suoi di cosa pertella." In a pertinent reflection Rigolot (2008, 94–95) studies the creative agency of willful erroneousness in Du Bellay and other poets.

8. For Alfred Glauser (1967, 24) the hare is the poet's totem. It reflects a "metaphysical disquiet . . . equal to the poet's disquiet." Scève's hare also has a comparative analogue in the *Hécatomgraphie,* "Peril & danger de tous costez" (emblem 47),

"Comme ce liepvre est pris de touts costez / Et n'a refuge en terre ne en mer, / En tous perilz ainsi sommes boutez . . ." (fol. G iv r). [Just as this hare is surrounded on all sides / and has refuge neither on land or sea / so then do we confront every peril . . .]

9. The difference and repetition of the forms point to a quandary of origin and error: Does the printer have only one style of vertical ovals available? Or does its rich grotesque design make the cartouche especially appealing (as shown later in this chapter)? Both questions can be answered in the affirmative.

10. The beauty of the design was not lost on the editors of Defaux's critical edition. It figures on the cover of the second volume and in fact is far more arresting than the image of the "mirror that sees itself" chosen to herald the first volume.

11. The spider and its web also carry reminiscence of Pieter Apian's polar projection of the world, *rethe sive aranea nominatur* as it was described in the *Cosmographia* and shown as the base of a volvelle, the instrument with which individuals locate themselves in the world (chapter 3).

12. Defaux's extensive gloss of the poet-Prometheus is taken up in 2004 (1, lxvii–ci).

13. Illustrated in Cachey (2007a, 193).

14. Defaux alertly notes how Clément Marot might have been the "source" for the figure of fluvial inversion in an eclogue written for Francis I in 1539 (2004: 2, 4), in which "Plus tost le Rosne encontremont courra . . ." (No sooner than the Rhône will flow upward . . ."). Scève is shown speaking to Marot through the figure (1:xlv). And the use of *desassemble* (line 1) finds its origin in the *Adolescence clémentine* and the poet's translation (in 1541) of *L'histoire de Leandre et de Hero*. Defaux writes, "In this properly founding moment the landscape is not an origin; it is a metaphor, *a visual projection* and the inscription of the certitude that inhabits, conveys, and defines the subject" (2004: 1, xlvi–xlvii; emphasis added).

15. Defaux asserts that the dizain "must be read as a scenario of the erotic imaginary [or unconscious] of the Lover." Without further describing the imaginary he relates its qualities to those found in Rimbaud's "Bateau ivre": "O que ma quille éclate, ô que j'aille à la mer . . ." (D2, lines 247–48). [May my keel break, O that I go off to the sea.] Rimbaud is more pertinent than the editor suggests: the *enfant terrible* puns on the *quille* as keel and as his *quill,* the pen that writes the event and so abolishes what it simultaneously creates.

16. As in Maurice Bouguereau's *Théâtre françoys* (1594, fol. xiv v).

17. [Maurice Scève], *Saulsaye* (1547), fol. 8v, lines 109–21. Noteworthy is that Scève appears only in the name of his device at the end of the poem. Giordano (2010, 437–47) offers a rich study of dizain 346 in which the nuptials of the rivers Saône and Rhône are cast in the same way. When they conjoin in the sea, the two bodies of water undergo a "spiritual betrothal," which the Flemish mystic Jan van Ruusbroec took up in his writings on meditation. Giordano's work on *Délie* applies here.

18. Peter Sharratt (2005) places a highly enlarged reproduction of the image between the covers of his *Bernard Salomon: Illustrateur lyonnais,* as if to attest to Salomon's artistry. Sharratt's catalogue and history of the artist and his work are a point of reference for what follows.

19. It may be that the black marks are not only birds but also supports that assure a flush and even imprint on the paper onto which the image is transferred. The signs of birds could be a function of a technology of representation.

20. A variant can be found in the detail of the surveyor seen charting the image of Limania d'Auvergne in which he is found, in Gabriele Symeone's historical and topographical map of the region of 1561 (referenced in Karrow 1993, 527–28, and illustrated in Conley 1996, 234). The pedagogical quality of the scene is taken up in Lestringant (1993, 73–75).

21. Verdun-L. Saulnier (1948–49, 2:213) opts for the site of the Faculty of Letters. Sharratt (2005, 232) notes as well that a gap opens between the topographical "reality" the image represents and its more decisive role in the context of the poem.

22. Rosolato (1996, 153–54) has noted that a "known unknown" is felt with greater solitude than its dangerous counterpart, the "unknown unknown." The one and the other inhabit Scève's landscape.

5. Ronsard in Conflict

1. The writing of Ronsard's life (1524–85) runs from the *Odes* of 1550 to the *Derniers vers* that the author composed as he lay dying. The principal albeit incomplete study of the alterations is Terreaux (1968). Dassonville's biography ([1975]–84) remains a point of reference for what concerns the life and times of the author. The complex history of the *Discours,* as an end point to this chapter, is carefully studied in Barbier (1984).

2. Ronsard, *Les Amours* (1552), fol. 31r; reprinted in *Les amours,* ed. Henri Weber and Catherine Weber, 41. Reference is made to the *editio princeps* to capture the force of the poetry before variants in the later editions tend to "correct" and control its measure. This volume is based on the critical edition, given in the *editio princeps,* begun by Paul Laumonier and completed by Isidore Silver and others (twenty volumes, 1914–75). Laumonier, a cornerstone, will be designated by an *L,* followed by the volume and page number. The final edition (1584) published during Ronsard's life is the basis for the *Oeuvres complètes,* edited by Jean Céard, Daniel Ménager, and Michel Simonin (1994). It will be designated by *PL,* followed by the volume and page number. Following the aspect of the first edition, the poems are shown here in italics. The *speed* with which the poems are seen and read is made clear where the printed characters seem to bend with the wind of anticipation. Noteworthy, too, is that in the early editions of the *Amours* and their continuation, the visual aspect of the sonnet is never broken. Two poems occupy each side of the folio. Modern editions take little care in the presentation of the sonnet as such.

3. Bartolomeo Sonetti's *Isolario* of 1485 juxtaposes sonnets describing the Aegean archipelago to a map of each island encountered and described. The space and jagged outline of the map, a point of reference along the poet's journal and a topic of the poem, cause the cartographic image and the text to inflect each other. Hence the latent form of the sonnet as an island. Elsewhere (Conley 1992/2006, chapters 3–4) study is made of sonnets 59 and 90 from this angle.

4. Ronsard, *Les amours* (1552), fol. 33r. See also Christine de Buzon and Pierre Martin (1999), 99 (1553 edition, 79), whose orthography is somewhat different. The orthography in the 1560 edition of *Les oeuvres de P. de Ronsard* (1560), 1, 42v, is close to that of the 1553 edition. The Weber and Weber edition of 1963/1998 follows the orthography

of 1552. In a lapsus it prints *et* in place of *&* in line 7: "Coustaux vineus, et plages blondoyantes" (43).

5. In a careful and enlightened reading of Ronsard's rewriting of Bevilacqua's poem and the commentary it has inspired, JoAnn DellaNeva (2010, chapter 7) appears at odds with Donald Stone's conclusion that where the Italian model offers well-ordered apostrophes, of which each "'forms a separate thought and is enclosed within a natural division in the sonnet,'" in Ronsard "'there is no such integration of elements, but simply the interminable list.'" (This reader counts a musical enumeration and regathering of twelve elements that convey the sensation of the landscape.) DellaNeva is closer to Henri Weber (1955, 321), who notes that the detailed imagery of the landscape "is tied to [the poet's] love of the Vendômois countryside" and that "the impetuous vivacity of the rhythm expresses the shudder of an unusually mobile sensitivity." With allusion to Louisa MacKenzie (2002), DellaNeva remarks how the words *cousteaux vineux, & plages blondoyantes* turn Bevilacqua's "generic landscape" into one of his own *terroir*.

6. In "Les isles fortuneés" (1555), Ronsard imagines a virgin land where, "Sans tailler la nourrissiere plante / Du bon Denys, d'une grimpeure lente / S'entortillant meurira ses raisins / De son bon gré sur les ormes voisins" (PL, 2, 782, lines 85-88) [Without pruning, great Dionysus's nourishing plant / gripping slowly as it goes, / winding about, will willfully ripen its grapes / on the neighboring elms]. In a reading that bears indirectly on the "Discours contre Fortune" (taken up later in this chapter), Daniel Ménager (2008, 21–35) sets the landscape of the "old" world in view of that of the "new."

7. Gilbert Gadoffre (1960/1994, 98–99 and 120) notes this point indirectly, by way of illustration of details from the painter's *Beating of Wheat* and *Sifting of Seeds* (in the Louvre).

8. The effect is cinematic if Gilles Deleuze's notion of paratactic addition is brought forward (1985, 241–42). For Deleuze Jean-Luc Godard *adds* images one after the other so as to make a moving collage. Comparison with Ronsard is not unworthy, especially in the rapid acceleration that ends in balance and measure in the sight and sound of the last word: *moy*.

9. The myth in the late 1550s of Ulysses as the redemptive traveler, he who led a "beau voyage," is taken up in Gadoffre (1964, 122–23 and 210–12). Thevet's mix of real and imaginary travel is taken up in Lestringant (2003, 136–44 and passim).

10. Ronsard had already (in 1559) written of his acquaintance with Nicolas de Nicolaï's topographical map of the Boullonnais (which would later appear in Ortelius's *Theatrum orbis terrarum,* 1570) and Maurice Bouguereau's *Théâtre françoys* (1594, with which he would further underscore the presence of the world in tiny traits). He gave Montmorency a copy of the map that attests to Henry II's campaign in the northwest. On the map is the notation: "Vous y verrez Calais au naturel dépeint" (line 5) [there you'll see Calais depicted as it is]—and further, "Vous verrez la grandeur, les places, & les forts / Du Boulongnois, & d'Oye, & la mer, & les ports, / Monts, fleuves, & forests, qui s'ejouissent d'estre / Reduits de sous la main de leur ancien maistre" (lines 8–11). [You'll see the grandeur, the sites, the forts / Of the Boulonnais and the Oye, the sea, and the ports, / Mountains, rivers, and forests, happily / taken away from their former master.] The enumeration is indeed faithful to what is shown on the map.

In the greater context of the extension and expansion of vision that emerge from the *Second livre* the words implicitly ask the reader to look at the surrounding poems in the same way.

11. In his notes on these lines Laumonier insists on the allusion to Ronsard's reconciliation with Saint-Gelais and his recognition of Marot and his disciples. The lines tell more about where he is in the order of things, in a universal history.

12. Claude Lévi-Strauss, in Charbonnier (1969, 33). The cold society is to the modern or hot counterpart as a clock is in relation to a steam engine. The former creates a minimum of entropy and remains in its initial state.

13. In *Tristes tropiques* (1955), first recalling Montaigne's tale, in "Des cannibales," of his fabled visit to Rouen, Lévi-Strauss shows that traditional cultures build their social bonds on the foundation of voluntary servitude: a compact invests its faith in its leader; it expends its energies to honor the labors of the leader; if the latter does not obtain his or her power through generosity, the former have the right to rebel (356–59).

14. And he will soon deploy an encomium of the snail as well. In the *Response aux injures et calomnies de je ne sçay quels predicantereaux et ministreaux de Genève* (1563), he compares himself to the animal that carries its house with it wherever it goes. The poet, notes Anne-Pascale Pouey-Mounou (2002, 274), calls the snail a "garden warrior" whose "round house" is seen as a celebration of the cosmos. She infers that Rémy Belleau's snail, in the *Petites inventions* (in Schmidt 1953, 538–42), belongs to the tradition of the great praise of small things. Belleau's and Ronsard's encomia anticipate those of Peter Williams's *Snail* (2008).

6. Montaigne and His Swallows

1. The pre-Cartesian pleasure of mapping a new place on a flat surface is present in the Latin title: "De descriptione regionis alicuius in *plano*, incognitis latitudine, longitud. & distancia" (Apian 1553, fol. 59v; emphasis added). The Latin title of the appendix, *Libellus de Locoroum describendorum ratione, Et de eorum distantiis inveniendis, nunquam antehac visus* (fol. 59r), is also indicative. To find, *trouver*, has in *invenire* the inference of "happening upon" and of "inventing" in the sense discovered in Rabelais and among the poets studied in this volume.

2. *Cosmographie* (1553a), fol. 52v. The Latin reads thus: "Igitur si nunc Provinciam totam depingere placet, investiga primum ab uno Oppido, à quo placet incipere, omnium circumiacentium locorum situs, eósque trahe in plano descripto primum circulo ex uno puncto posito ad libitum. . . . Ut autem evites multam peregrinationem, ascende turrim Oppidi altissimam, atque inde quasi à specula cirumspice. Post haec proficiscere ad aliud oppidum, atque ibi familiter agito cum angulis positionum omnium circumiacentium locorum" (Paris: V. Gaulthrot, 1553), fol. 61r.

3. The topography of the writing space is taken up at the end of "De trois commerces." Space and meditation in Montaigne's tower are taken up in Conley (2009, 95–98).

4. It is to be wondered whether Montaigne's reference to gutters concerns the leaden conduits under the eaves of the house or his urinary tract blocked by kidney stones. "*Stilicidi casus lapidem cavat. Ces ordinaires goutieres me mangent. Les inconvenients*

ordinaires ne sont jamais legiers. Ils sont continuels & irreparables, nomméement quand ils naissent des membres du mesnage, continuels & inseparables" (928/949). [*The droplet of water pierces the stone.* These ordinary gutters eat at me. Ordinary difficulties are never light. They are continual and irreparable, namely, when they are born of the members of the household, continual and inseparable.] In this movement of "De la vanité" he avows his "insuffisance aux occupations du mesnage" (929/950) [insufficiency in household occupations], noting that he would prefer to be "bon escuyer que bon logitien" (ibid.) [a good lackey than a good logician]. In using *logitien* the author puns on logic and *logis,* the lodging or habitus where one lives, and also the place affiliated with topographers—*maîtres de logis*—in the omnipresent military context of the Wars of Religion. See note 5 regarding references to the *Essais.*

5. As in the Introduction, references to Montaigne's *Essais* are to the 1962 edition of the *Oeuvres complètes* (Rat and Thibaudet) and the 1988 edition of the *Essais* (Villey and Saulnier), respectively.

6. The opening page of the essay invokes an aura of knowledge in the foyer of his father's house ("ma maison a esté de long temps ouverte aux gens de sçavoir, et en est fort connuë: car mon pere, qui l'a commandée cinquante ans et plus, eschauffé de cette ardeur nouvelle dequoy le Roy François Ier embrasse les lettres et les mis en credit, rechercha avec grand soing et accointance des hommes doctes" [438–39/481]), but under the superscription of the first sentence in which science and letters are praised, is cast the shadow of an uncanny bestiality ("C'est, à la verité, une tres-utile et grande partie que la science, ceux qui la mesprisent, tesmoignent assez de leur bestise" [438/481]).

7. Especially Donald M. Frame, in *Montaigne's Discovery of Man: The Humanization of a Humanist* (1955), who espouses Villey's view and who reflects it to some degree in his magnificent translation of the *Essays* (1958).

8. Introjection is used in concert with that of Rabelais, as it was explored in chapter 1. Montaigne introjects received knowledge slyly and with ruse. The form of the essay shares much with Jacques Derrida's brilliant performance of introjection in *Limited Inc abc . . .* (1977), an essay in response to John Searle in which Derrida used a lawyer's tactic of *chicane* or double-talk in which, in flagrant violation of copyright laws, he quoted the entirety of his adversary's essay that had called one of the philosopher's previous pieces (on speech-acts) into question.

9. Michel Foucault uses form of content to describe the gridding of penal orders in post-1789 France in *Surveiller et punir* (1975). In "Un nouveau cartographe?" in *Foucault* (1986, 49–51), Gilles Deleuze extends the field of meaning to include discursive and visible formations in general.

10. In *Montaigne paradoxal* (1970), Alfred Glauser intuits that the essays create themselves by virtue of paradox. Here it can be added that in its relation to cosmography the physical shape of the essay adduces the same process. It is both infinitizing and extensive in movement, but also alerts the reader to a nonspace that is always outside the confines it is building.

11. In another context it could be argued that in this and other passages the form of the content of the "Apologie" marks a decisive moment in the political history of ecology and philosophy. The world of animals, romantic philosophers would want to believe, is not expansive or prone to destruction of ecosystems. Where they are in homeostasis their self-corrective mechanisms serve as models with which to contrast the

"runaway" conditions that human beings have imposed upon the environment. Gregory Bateson studies the relation of dynamic and chaotic systems in what he fittingly calls "Pathologies of Epistemology" (1972, 478–87); Deleuze and Guattari (1991) show how birds in fact use art to fashion their relation with the world. The "Apologie" *stages* the effects of a system that veers toward a "runaway" condition all the while it constitutes a decisive critique of what happens when "man" uses reason to exploit the *oikos*.

12. Gerard Brett (1954, 477–87), Derek J. de Solla Price (1975), and William Eamon (1983) call attention to nature's machines. I am grateful to Denyse Delcourt for having brought these and other studies to my attention.

13. Clément Marot, *Oeuvres complètes,* vol. 1 (2007).

14. By "naturalistic" is meant what Henri Meschonnic (1994) describes as a moving, mobile relation (and not a schematic analogy) between a graphic form and its referent. Montaigne's "arondelle" would be no longer "a movement that originates ahistorically in nature and motivates language. It is a discursive act, *nature in voice,* much as 'words' in Aristotle's *De interpretatione* are . . . things that are in the voice" (103). The bird thus moves, it tracks an itinerary, it translates itself in a world of both infinite and finite grandeur.

15. When referring to the halcyon, Belon keeps the kingfisher in mind, a bird that makes nests somewhat comparable to those of the fabled creature in the "Apologie." "Quelque part qu'on trouve le nid d'un martinet pescheur, lon doit penser que celuy d'un Hirondelle de rivage, n'en est moult loing: & de vray sçachants que son bec est foible, & petit, pensons qu'elle ne creuse la terre pour le faire: mais qu'elle entre en celuy des Halcyons, ou martinents pescheurs, esquels avoient nourry leurs petits l'année precedente. Car l'Halcyon est coustumier de faire un nouveau creux par chaque annee, entendu qu'il a le fort bec, long et dur" (fol. 379). [Wherever the kingfisher's nest is found it must be thought that that of a shoreline swallow is not very long; and indeed, knowing that its beak is small and weak, we think that it does not dig about the earth to do it: but that it enters into that of the Halcyons or kingfishers who have nourished their little ones the year before. For the Halcyon customarily makes a new hollow every year for the reason that it has a strong, long and hard beak.]

16. Lestringant (2002 and 2003) situates Thevet as the exemplar par excellence of the textual isolario in Renaissance France.

17. Deleuze (1985, 214). The concept comes on the heels of the formulation of an event that resembles the spatial effect of the "Apologie." "Reste que, dans le devenir, la terre a perdu tout centre, non seulement en elle-même, mais elle n'a plus de centre autour duquel tourner. Les corps n'ont plus de centre, sauf celui de leur mort quand ils sont épuisés, et rejoignent la terre pour s'y dissoudre" (186). [It remains that in becoming, the earth has lost every center, not only in itself, but it no longer has a center around which to turn. Bodies no longer have a center, except that of their death when they are exhausted and return to the earth to dissolve themselves within it.]

18. In *Pré-histoires 1* (1999), Terence Cave offers an alternate reading of the insertion in the "Apologie," in which it is shown that Montaigne discerns a prescient anthropology in the poem for the reason that Ronsard assumes the point of view of the pagan.

Bibliography

Abraham, Nicolas, and Maria Torok. 1978. *L'écorce et le noyau*. Paris: Aubier-Flammarion.

Adams, Alison, Stephen Rawles, and Alison Saunders, eds. 1999 and 2002. *A Bibliography of French Emblem Books*. 2 vols. Geneva: Droz.

Alciati, Andreas. 1534. *Liber emblematum*. Paris: C. Wechelus. [Houghton Library, Harvard University Typ 515.34.132]

Alduy, Cécile. 2001. "Introduction" to her reedition of Maurice Scève, *Délie: Obiect de plus haute vertu*, v–xlvi. Paris: STFM.

———. 2003. "'Délie' palimpseste ou l'art de la citation." *Studi francesi* 47, no. 1: 23–38.

———. 2007. *Politique des "Amours": Poétique et genèse d'un genre français nouveau (1544–1560)*. Geneva: Droz.

Alpers, Svetlana. 1983. *The Art of Describing: Dutch Art in the Seventeenth Century*. Chicago: University of Chicago Press.

Apian, Pieter. 1532. *Cosmographiae introductio: Cum quibisdam geometriae astronomiae principiis ad eam rem necessarii*. Ingolstadt: n.p.

———. 1535. *Cosmographiae introductio . . .* Venice: Antonium de Nicolinis de Sabio.

———. 1544. *La cosmographie de Pierre Apian, libvre tresutile, traictant de toutes les regions & pays du monde par artifice Astronomicque, nouvellement traduict de Latin en François. Et par Gemma Frison Mathematicien & Docteur en Medicine de Louvain corrigé. Avecq aultres libvres du mesme Gemma Frisium . . .* Antwerp: Gregoire Bonte.

———. 1550. *Cosmographia Petri Apiani, per Gemmam Frisium apud Louanienses Medicum & Mathematicum insignem, iam demum ab omnibus vindicata mendis, ac nonnullis quoque locis aucta, figurisque novis illustrata: Additis eiusdem argumenti libellis ipsius Gemmae Frisii*. Antwerp: Gregorio Bontio.

————. 1551. *La cosmographie de Pierre Apian, docteur et mathematicien tres excellent, traictant de toutes les Regions, Pais, Villes, & Citez du monde, Par artifice Astronomique, nouvellement traduicte de Latin en François par Gemma Frisius, Docteur en Medecine, & Mathematicien de l'université de Louvain, de nouveau augmentée, oultre les precedentes impressions, comme l'on pourra veoir en la page suyvante. Le tout avec figures a ce convenantes, pour donner plus facile intelligence.* A Paris: Par Vivant Gaultherot, libraire juré en l'université de Paris, demourant en la rue de Sainct Iaques, a l'enseigne S. Martin.

————. 1553. *Cosmographia Petri Apiani, per Gemmam Frisium apud Louanienses Medicum & Mathematicum insignem . . .* Paris: vaeneunt apud Vivantium Gaultherot, via Iacobea: sub intersignio D. Martini.

Artaud, Antonin. 1925. *L'ombilic des limbes.* Paris: Editions de la Nouvelle Revue Française. [Houghton Library, Harvard University FC9.Ar752.925o]

Auerbach, Erich. 1953. *Mimesis: The Representation of Reality in Western Literature.* Translated by Willard R. Trask. New York: Doubleday.

Augé, Marc. 1992. *Non-lieux: Introduction à une anthropologie de la surmodernité.* Paris: Seuil.

Bagnoli, Martina, ed. 2009. *Prayers in Code: Books of Hours from Renaissance France.* Baltimore: The Walters Art Museum.

Baraz, Michaël. 1983. *Rabelais et la joie de la liberté.* Paris: José Corti.

Barbier, Jean-Paul. 1984. *Bibliographie des discours politiques de Ronsard.* Geneva: Droz.

Bateson, Gregory. 1972. *Steps to an Ecology of Mind.* New York: Ballantine Books.

Belon, Pierre. 1555. *L'histoire de la nature des oyseaux.* Paris: B. Prévost/Gilles Corrozet. [Houghton Library, Harvard University Typ 515.55.201]

Béroalde de Verville. 1612?/2006. *Le moyen de parvenir.* Edited by Michel Jeanneret. Paris: Gallimard/Folio Classique.

Bertrand, Antoine de. 1587. *Premier livre des "Amours" de Pierre de Ronsard mis en musique à iiii. parties.* Paris: Adrien le Roy and Robert Ballard. [Houghton Library, Harvard University FC5.R6697.Gz576bc]

Besse, Jean-Marc. 2003. *Les grandeurs de la terre: Aspects du savoir géographique à la Renaissance.* Paris: ENS.

Blanchot, Maurice. 1969. *L'entretien infini.* Paris: Gallimard.

————. 1973. *Le pas au-delà.* Paris: Gallimard.

Bouguereau, Maurice. 1594. *Le théâtre françoys.* Tours: Maurice Bouguereau.

Braun, Georg. 1574–1618. *Théâtre des cités du monde.* 6 vols. Brussels: n.p.

Brett, Gerard. 1954. "The Automata in the Byzantine 'Throne of Soloman.'" *Speculum* 28:477–87.

Brun, Robert. 1930. *Le livre illustré en France au XVIe siècle.* Paris: Alcan.

Buisseret, David. 2003. *The Mapmaker's Quest: Depicting New Worlds in Renaissance Europe.* Oxford: Oxford University Press.

Butsch, Albert. 1878/1969. *Handbook of Renaissance Ornament: 1290 Designs from Decorated Books.* Reprint, New York: Dover Books.

Buzon, Christine de, and Pierre Martin, eds. 1991. Critical edition of Pierre de Ronsard, *Les "Amours."* Paris: Classiques Didier Érudition.

Cachey, J. Theodore, Jr., ed. 2007a. *The First Voyage around the World, 1519–1522:*

An Account of Magellan's Expedition by Antonio Pigafetta. Toronto: University of Toronto Press.

———. 2007b. "Maps and Italian Literature of the Renaissance." In Woodward 2007, 182–90.

Cave, Terence. 1984. *The Cornucopian Text: Problems in Writing in the French Renaissance.* Oxford: Clarendon Press.

———. 1999. *Pré-histoires: Textes troubles au seuil de la modernité.* Geneva: Droz.

Céard, Jean, and Jean-Claude Margolin. 1984. *Rébus de la Renaissance: Des images qui parlent.* 2 vols. Paris: Maisonneuve et Larose.

Certeau, Michel de. 1975. *L'écriture de l'histoire.* Paris: Gallimard/Bibliothèque des Histoires.

———. 1982. *La fable mystique: XVIe–XVIIe siècles.* Paris: Gallimard/Bibliothèque des Histoires.

———. 1984/1996. "Entretien avec Alain Charbonnier et Joël Magny." *Ça Cinéma* no. 301 (January 1984): 19–20. Reprinted in "Feux persistants: Entretien sur Michel de Certeau." *Esprit* no. 219 (March 1996): 148.

———. 1990. *L'invention du quotidien.* Vol. 1, *Arts de faire.* Edited by Luce Giard. Paris: Gallimard/Folio.

Charbonnier, Georges. 1961/1969. *Conversations with Claude Lévi-Strauss.* Translated by John Weightman and Doreen Weightman. London: Jonathan Cape.

Charpentier, Françoise. 1984. "En si durs epigrammes." Preface to her edition of Maurice Scève, *Délie,* 7–46. Paris: Gallimard/Poésie.

Chatelain, Jean-Marc, and Laurent Pinon. 1999. "Genre et fonctions de l'illustration au XVIe siècle." In Martin 1999, 236–69.

Colonna, Francesco. 1546. *Hypnerotomachie; ou, Discours du songe de Poliphile . . .* Translated by Jan Martin. Paris: Jacques Kerver. [Houghton Library, Harvard University Typ 515.46.296 F]

Conley, Tom. 1992. *The Graphic Unconscious in Early Modern French Writing.* Cambridge: Cambridge University Press.

———. 1996a. *The Self-Made Map: Cartographic Writing in Early Modern France.* Minneapolis: University of Minnesota Press.

———. 1996b. "Holbein's Lacan: The Wit of the Letter." In *Vision in Context: Historical and Contemporary Perspectives on Sight,* edited by Teresa Brennan and Martin Jay, 45–61. London: Routledge.

———. 2002. "Montaigne *moqueur:* 'Virgile' and Its Geographies of Gender." In *High Anxiety: Masculinity in Crisis in Early Modern France,* edited by Kathleen Perry Long, 93–106. Kirksville, MO: Truman State University Press.

———. 2006. "An Eclogue Engraved: Maurice Scève and Bernard Salomon's *Saulsaye* (1547)." In *Book and Text in France: Poetry and the Page,* edited by Adrian Armstrong and Malcolm Quainton, 139–62. Aldershot: Ashgate Press.

———. 2007. *Cartographic Cinema.* Minneapolis: University of Minnesota Press.

———. 2009. "Le mediter: Via Apian." *Cahiers Verdun-L. Sulnier* no. 26: 95–112.

Corrozet, Gilles. 1534. *La fleur des antiquitez, singularitez, & excellences de la noble & triumphante ville & cité de Paris, capitale du royaulme de France, adioustees oultre la premiere impresession plusieurs singularitez estans en ladicte ville. Auec la genealogie du roy Francoys premier de ce nom.* Paris: Galiot du Pré.

————. 1538. *Les simulachres & histories faces de la mort, autant elegamment pourtraictes que artificiellement imaginées.* With woodcuts by Hans Holbein. Lyons: Soubz l'escu de Coloigne. [Houghton Library, Harvard University Typ 38.456]

————. 1539. *Les blasons domestiques, contenantz la decoration d'une maison honneste, & du mesnage estant en icelle: Invention joyeuse, & moderne.* Paris: [Denys Janot] for Gilles Corrozet. [Houghton Library, Harvard University Typ 515.39. 299]

————. 1541. *Hécatomgraphie: C'est à dire les descriptions de cent figures & hystoires, contenans plusieurs appothegmes, proverbes, sentences & dictz tants des Anciens que des modernes . . .* Paris: Denys Janot.

————, trans. 1542. *Les fables du tres ancien Esope phrigien premierement escriptes en Graec, & depuis mise en Rithme Françoise.* Paris: Denys Janot. [Houghton Library, Harvard University Typ 515.42.123]

————. 1544/1997. *Hécatomgraphie, 1544; & Les emblèmes du tableau de Cèbes, 1543.* Facsimile and critical edition by Alison Adams. Geneva: Droz.

————, trans. 1547. *Les fables d'Esope phrygien, mises en Ryme Françoise: Avec la vie dudit Esope extraite de plusieurs autheurs par M. Antoine du Moulin Masconnois.* Lyons: Jean de Tournes and Guillaume Gazeau. [Houghton Library, Harvard University Typ 515.47.123]

————. 1586. *Les antiquitez, croniques et singularitez de Paris, ville capitalle du royaume de France . . .* 2 books in 1 vol. Paris: Nicolas Bonfons.

Cosgrove, Denis. 2001. *Apollo's Eye: A Cartographic Genealogy of the Earth in the Western Imagination.* Baltimore: The Johns Hopkins University Press.

————. 2007. "Images of Renaissance Cosmography, 1450–1650." In Woodward 2007, 55–98.

Cotgrave, Randle. 1611. *Dictionarie of the French and English Tongues.* London: Adam Islip.

Dainville, S.J. François de. 1964/2002. *Le langage des géographes.* Paris: A. et J. Picard.

————. 1970. "How Did Oronce Finé Draw His Large Map of France?" *Imago Mundi* 24: 49–55.

Dassonville, Michel. 1975–84. *Ronsard: Étude historique et littéraire.* 4 vols. Geneva: Droz.

Defaux, Gérard, ed. 2003. *Lyon et l'illustration de la langue française à la Renaissance.* Lyon: Editions ENS.

————, ed. 2004. Critical edition of Maurice Scève, *Délie* (1544). 2 vols. Geneva: Droz (Coll. Textes Littéraires Français).

Dekker, Ely. 2007. "Globes in Renaissance Europe." In Woodward 2007, 135–59.

Delano-Smith, Catherine. 2006. "Milieus of Mobility: Itineraries, Route Maps, and Road Maps." In *Cartographies of Travel and Navigation,* edited by James Akerman, 16–68. Chicago: University of Chicago Press.

————. 2007. "Signs on Printed Topographical Maps, ca. 470–1640." In Woodward 2007, 528–90.

Deleuze, Gilles. 1983. *Cinéma 1: L'image-mouvement.* Paris: Minuit.

————. 1985. *Cinéma 2: L'image-temps.* Paris: Minuit.

————. 1986. *Foucault.* Paris: Minuit.

————. 1988. *Le pli: Leibniz et le baroque.* Paris: Minuit.

————. 1992. "L'epuisé." In Samuel Beckett, *Quad.* Paris: Minuit.

Deleuze, Gilles, and Félix Guattari. 1980. *Capitalisme et schizophrénie*. Vol. 2, *Mille plateaux*. Paris: Minuit.

———. 1991. *Qu'est-ce que la philosophie*. Paris: Minuit.

DellaNeva, JoAnn. 2010. *Unlikely Exemplars: Reading and Imitating beyond the Italian Canon in Renaissance France*. Newark: University of Delaware Press.

Derrida, Jacques. 1977. *Limited Inc: abc. . .* Baltimore: The Johns Hopkins University Press.

Desan, Philippe, ed. 2002. *Color Reproduction of the Bordeaux Copy of Montaigne's "Essais."* Chicago: University of Chicago, Montaigne Studies.

———, ed. 2006. *Dictionnaire Montaigne*. Paris: Champion.

Du Bellay, Joachim. 1549/2001. *La deffence et illustration de la langue françoyse*. Critical edition and dossier, edited by Jean-Charles Montferran. Geneva: Droz.

———. 1554. *L'olive augmentee depuis la premiere edition: La musagnoemachie & autres oeuvres poetiques*. Paris: Gilles Corrozet & Arnoul L'Angelier.

———. 1558. *Les regrets et autres oeuvres poetiques de Ioach. Du Bellay, ang.* Paris: Federic Morel.

Dunlop, Anne. 2009. *Painted Palaces: The Rise of Secular Art in Early Renaissance Italy*. University Park: Pennsylvania State University Press.

Duval, Edwin. 1991. *The Design of Rabelais's "Pantagruel."* New Haven, Conn.: Yale University Press.

———. 1998. *The Design of Rabelais's "Quart livre de Pantagruel."* Geneva: Droz.

Eamon, William. 1983. "Technology and Machine in the Middle Ages and Renaissance." *Janus* 70, no. 3–4: 170–203.

Edgerton, Samuel Y., Jr. 1987. "From Mental Matrix to Christian Empire: The Heritage of Ptolemaic Cartography in the Renaissance." In *Art and Cartography: Six Historical Essays*, edited by David Woodward, 10–50. Chicago: University of Chicago Press.

Edson, Evelyn. 2007, *The World Map 1300–1492: The Persistence of Tradition and Transformation*. Baltimore: The Johns Hopkins University Press.

Eisenstein, Sergei. 1959. *The Film Form and the Film Sense*. Translated by Jay Leyda. New York: Meridian.

Finé, Oronce. 1542. *De mundi sphaera, sive Cosmographia . . .* Paris: Simon de Colines.

———. 1549. *Le sphere du monde: Proprement dicte Cosmographie*. Paris (manuscript with dedication to Henri II of France). [Houghton Library, Harvard University MS. Typ 57]

———. 1551. *Le sphere du monde, proprement ditte, Cosmographie: Composée nouvellement en françois & divisée en cinq livres: Comprenans la premiere partie de l'astronomie, & les principes universels de la geographie & hydrographie: avec une épistre, touchant la dignité, perfection & utilité des sciences mathématiques*. Paris: Michel de Vascosan.

Fiorani, Francesca. 2005. *The Marvel of Maps: Art, Cartography, and Politics in Renaissance Italy*. New Haven, Conn.: Yale University Press.

Focard, Jacques. 1546. *Paraphrase de l'astrolabe*. Lyons: Jean de Tournes.

Foucault, Michel. 1975. *Surveiller et punir*. Paris: Gallimard/Bibliothèque des Histoires.

Foullon, Albert. 1561. *Vsaige et description de l'holomètre: Pour sçavoir mesurer toutes choses qui sont soubs l'estanduë de l'oeil: Tant en longhueur & largeur, qu'en hauteur & profondité.* Paris: Pierre Beguin.

Frame, Donald M. 1955. *Montaigne's Discovery of Man: The Humanization of a Humanist.* New York: Columbia University Press.

———, trans. 1958. *Montaigne's Complete Essays.* Stanford, Calif.: Stanford University Press.

———. 1977. *Rabelais: A Study.* New York: Harcourt Brace Jovanovich.

Freud, Sigmund. 1957/1986. *The Standard Edition of the Complete Psychological Works of Sigmund Freud.* Edited by James Strachey. 24 vols. London: The Hogarth Press.

Gadoffre, Gilbert. 1960/1994. *Ronsard.* Paris: Seuil, Coll. "Écrivains de Toujours."

———. 1978. *Du Bellay et la sacré.* Paris: Gallimard.

Gaudio, Michael. 2008. *Engraving the Savage: The New World and Techniques of Civilization.* Minneapolis: University of Minnesota Press.

Giordano, Michael J. 2010. *Meditation and the French Renaissance Love Lyric: The Poetics of Introspection in Maurice Scève's "Délie, object de plus haulte vertu" (1544).* Toronto: University of Toronto Press.

Glauser, Alfred. 1966. *Rabelais créateur.* Paris: Nizet.

———. 1967. *Le poème-symbole: De Scève à Valéry.* Paris: Nizet.

———. 1970. *Montaigne paradoxal.* Paris: Nizet.

———. 1975. "Montaigne et 'le roseau pensant.'" *Romanic Review* 66: 263–68.

Le grand calendrier et compost des bergers. 1541. Troyes: Jean Lecoq.

Greene, Roland. 1999. *Unrequited Conquests: Love and Empire in the Colonial Americas.* Chicago: University of Chicago Press.

Grynaeus, Simon. 1532. *Novum orbis regionum ac insularum veteribus incognitarum.* Paris: Apud Ioannum Paruum.

Guéroult, Guillaume. 1553. *Epitome de la corographie d'Europe, illustré des pourtraitz des villes plus renommées d'icelle.* Lyons: Balthazar Arnoullet.

Gundersheimer, Werner, ed. 1971. *The Dance of Death by Hans Holbein the Younger.* Facsimile of the original 1538 edition of *Les simulachres & histories faces de la mort.* New York: Dover.

Hallyn, Fernand. 2008. *Gemma Frisius, arpenteur de la terre et du ciel.* Paris: Champion.

Hampton, Timothy. 2001. *Literature and Nation: Inventing Renaissance France.* Ithaca, N.Y.: Cornell University Press.

Harvey, P. D. A. 1987. "Local and Regional Cartography in Medieval Europe." In *Cartography in Prehistoric, Ancient, and Medieval Europe and the Mediterranean.* Vol. 1, *The History of Cartography,* edited by David Woodward, 492–93. Chicago: University of Chicago Press.

Henkel, Arthur, and Albrecht Schöne, eds. 1967. *Emblemata: Handbuch zur Sinnbildkunst des XVI. und XVII. Jahrhunderts.* Stuttgart: J. B. Metlersche.

Hodges, Elisabeth. 2008. *Urban Poetics in the French Renaissance.* Aldershot, England, and Burlington, Vt.: Ashgate.

Horapollo. 1543. *De la signification des notes hieroglyphiques des Aegyptiens: C'est à dire des figures par les quelles ilz escripvoient leurs mystères secretz, et les choses sainctes et divines/Nouvellement traduict . . . [par Jean Martin].* Paris: Jacques Kerver. [Houghton Library, Harvard University Typ 515.43.455]

Isidore of Seville. 1473. *Etymologiae.* Augsburg: n.p.

Jacob, Christian. 2006. *The Sovereign Map: Theoretical Approaches in Cartography throughout History*. Edited by Edward Dahl and translated by Tom Conley. Chicago: University of Chicago Press.

Jacquinot, Dominicq. 1558. *L'usage de l'astrolabe avec un petit traicté de la sphere*. Paris: Guillaume Cavellat.

Jameson, Fredric. 1991. *Postmodernism; or, The Culture of Late Capitalism*. Durham, N.C.: Duke University Press.

———. 1992. *The Geopolitical Aesthetic: Cinema and Space in the World System*. Bloomington: Indiana University Press.

Karrow, Robert W., Jr. 1993. *Mapmakers of the Sixteenth Century and Their Maps*. Chicago: Speculum Orbis Press.

Kennedy, William. 2003. *The Site of Petrarchism: Early Modern National Sentiment in Italy, France, and England*. Baltimore: The Johns Hopkins University Press.

La Perrière, Guillaume de. [1539]. *Le theatre des bons engins, auquel sont contenuz cent emblemes moraulx*. Paris: Denys Janot. [Houghton Library, Harvard University Typ 515.39.511]

———. 1993. *Le théâtre des bons engins; La morosophie*. Edited by Alison Saunders. Aldershot, England, and Brookfield, Vt.: Scolar Press/Ashgate.

Lacan, Jacques. 1966. *Écrits*. Paris: Seuil.

Laplanche, Jean, and Jean-Baptiste Pontalis. 1976. *Vocabulaire de la psychanalyse*. Paris: Presses Universitaires de France.

Lavezzo, Kathy. 2006. *Angels on the Edge of the World: Geography, Literature, and the English Community, 1000–1534*. Ithaca, N.Y.: Cornell University Press.

Lecoq, Anne-Marie. 1987. *François premier imaginaire*. Paris: Macula.

Lefebvre, Henri. 1955. *Rabelais*. Paris: Français Réunis.

———. 1974. *La production de l'espace*. Paris: Editions Anthropos.

———. 1991. *The Production of Space*. Translated by Donald Nicholson-Smith. Oxford: Blackwell, 1991.

Lemaire de Belges, Jean. 1512–13. *Les illustrations de Gaule et singularitez de Troye . . . avec les deux epistres de lamant verd*. 6 parts in 1 vol. Paris: G. de Marnef.

Lester, Toby. 2009. *The Fourth Part of the World: The Race to the Ends of the Earth, and the Epic Story of the Map That Gave America Its Name*. New York: Free Press.

Lestringant, Frank. 1993. *Écrire le monde à la Renaissance*. Caen: Paradigme.

———. 2002. *Le livre des isles: Atlas et récits insulaires de la Genèse à Jules Verne*. Geneva: Droz.

———. 2003. *Sous la leçon des vents: Le monde d'André Thevet, cosmographe à la Renaissance*. Paris: Presses de l'Université de Paris-Sorbonne.

———. 2006. "Paysages anthropomorphiques à la Renaissance." In *Nature & paysages: L'émergence d'une nouvelle subjectivité à la Renaissance*, edited by Dominique de Courcelles, 261–79. Paris: École des Chartes.

Lestringant, Frank, and Monique Pelletier. 2007. "Maps and Descriptions of the World in Sixteenth-Century France." In Woodward 2007, 1463–79.

Lévi-Strauss, Claude. 1955. *Tristes tropiques*. Paris: Plon.

———. 1991. *Histoire de lynx*. Paris: Plon.

Lindberg, David. 1976. *Theories of Vision from Al-Kindi to Kepler*. Chicago: University of Chicago Press.

Lindgren, Uta. 2007. "Land Surveys, Instruments, and Practitioners in the Renaissance." In Woodward 2007, 477–508.

Lyotard, Jean-François. 1986. *Le postmoderne expliqué aux enfants*. Paris: Galilée.

MacKenzie, Louisa. 2002. "'Ce ne sont pas des bois': Poetry, Regionalism, and Loss in the Forest of Ronsard's Gâtine." *Journal of Medieval and Early Modern Studies* 32: 343–74.

Mallarmé, Stéphane. 1948. *Oeuvres complètes*. Edited by Henri Mondor. Paris: Gallimard/Pléiade.

Malraux, André. 1951. *Les voix du silence*. Paris: Gallimard.

Mangani, Giorgio. 1998. *Il "mondo" d'Abramo Ortelio: Misticismo, geografia e collezionismo nel Rinascimento dei Paesi Bassi*. Modena: Franco Cosimo Panini.

———. 2005. *Cartografia morale: Geografia, persuasione, identità*. Modena: Franco Cosimo Panini.

Marot, Clément. 2007 and 2008. *Oeuvres complètes*. Edited by François Rigolot. 2 vols. Paris: Garnier/Flammarion.

Martin, Henri-Jean, ed. 1999. *La naissance du livre moderne: Mise en page et mise en text du livre française, XIVe–XVIe siècles*. Paris: Editions du Cercle de la Librairie.

McKinley, Mary. 1996. *Les terrains vagues des "Essais": Itinéraires et intertextes*. Paris: Champion.

Ménager, Daniel. 2008. "Conflits et évasion dans 'Les isles fortunées.'" In *Writers in Conflict in Sixteenth-Century France: Essays in Honour of Malcolm Quainton*, edited by Elizabeth Vinestock and David Foster, 21–35. Durham: Durham University Press.

Merliers, Jean des. 1575. *La practique de geometrie descripte et demonstree*. Paris: Gilles Gorbin.

Meschonnic, Henri. 1994. "Claudel et l'hiéroglyphe ou l'ahité des choses." In *La pensée de l'image,* edited by Gisèle Mathieu-Castellani, 99–119. Paris: Presses de l'Université de Paris-VIII/Vincennes-à-Saint-Denis.

Metz, Christian. 1991. *L'énonciation impersonnelle; ou, Le site du film*. Paris: Meridiens/Klincksieck.

Miernowski, Jan. 1997. *Signes dissimilaires: La quête des noms divins dans la poésie française de la Renaissance*. Geneva: Droz.

Montaigne, Michel de. 1595. *Les Essais du de Michel seigneur de Montaigne. Ed. nouuelle trouuee après le deceds de l'autheur reueuë & augmentée par luy d'un tiers plus qu'aux precedentes impressions*. Paris: Chez Michel Sonnius.

———. 1962. *Oeuvres complètes*. Edited by Maurice Rat and Albert Thibaudet. Paris: Gallimard/Pléiade.

———. 1988. *Essais*. Edited and revised by Pierre Villey under the direction of Verdun-L. Saulnier. 3 vols. Paris: Presses Universitaires de France/Quadrige.

———. 1998. *Journal de voyage en Italie*. Edited by François Rigolot. Paris: Presses Universitaires de France.

Mortimer, Ruth. 1964. *Harvard College Library Department of Printing and Graphic Arts, Catalogue of Books and Manuscripts*. Part 1: *French 16th Century Books*. Cambridge, Mass.: Belknap Press of Harvard University Press.

Navarre, Marguerite de. 1547. *Marguerites de la Marguerite des princesses . . . , tres illustre royne de Navarre*. Lyons: Jean de Tournes.

Naya, Emmanuel. 2008. "La crise de la signification chez Rabelais." In *La Renaissance décentrée*, edited by Frédéric Tinguely, 175–92. Geneva: Droz.

Nelson, Robert S., ed. 2000. *Visuality Before and Beyond the Renaissance: Seeing as Others Saw*. Cambridge: Cambridge University Press.

Nuti, Lucia. 1995. "Le langage de la peinture dans la cartographie topographique." In *L'oeil du cartographe et la représentation géographique du Moyen Age à nos jours*, edited by Catherine Bousquet-Bressolier, 53–70. Paris: Comité des Travaux Historiques et Scientifiques.

Ollson, Gunnar. 2007. *Abysmal: A Critique of Cartographic Reason*. Chicago: University of Chicago Press.

Ortelius, Abraham. 1570. *Theatrum orbis terrarum*. Antwerp: Coppenium Diesth.

———. 1581. *Theatre de l'univers, contenant les cartes de tout le monde . . .* Antwerp: Christophe Plantin.

Ortroy, Fernand. 1901/1963. *Bibliographie de l'oeuvre de Pierre Apian*. Amsterdam: Meridian.

———. 1920. *Bio-bibliographie de l'oeuvre de Gemma Frisius, fondateur de l'école belge de géographie . . .* Brussels: M. Lamertin.

Padrón, Ricardo. 2007. "Mapping Imaginary Worlds." In *Maps: Finding Our Place in the World*, edited by James Akerman and Robert Karrow, 255–87. Chicago: University of Chicago Press.

Pastoureau, Mireille. 1984. *Les atlas français, XVIe–XVIIe siècles*. Paris: Bibliothèque nationale de France.

Pelletier, Monique. 2009. *De Ptolémée à la Guillotière (XVe–XVIe siècle): Des cartes pour la France pourquoi, comment?* Paris: CTHS Géographie 6.

Péricaud, Antoine. 1851. *Bibliographie lyonnaise du XVe siècle*. Lyons: Perrin.

Petrarca, Francesco. 1514. *Opera volgare*. Edited by Alessandro Vellutello. Venice.

———. 1544. *Il Petrarca*. Lyons: Jean de Tournes.

Philieul, Vasquin. 1548. *Laure d'Avignon: Au nom et adveu de la Royne Catharine de Medici, Royne de France. Extraict du poete florentin, Françoys Petrarque, et mis en françoys par Vasquin Philieul de Carpentras*. Paris: G. Gazeau.

Pinet, Antoine du. 1564. *Plantz, pourtraitz et descriptions de plusieurs villes et forteresses, tant de l'Europe, Asie, & Afrique, que des Indes, & terres neuves*. Lyons: Jan Ogerolles.

Pouey-Mounou, Anne-Pascale. 2002. *L'imaginaire cosmologique de Ronsard*. Geneva: Droz.

Price, Derek J. de Solla. 1975. *Science since Babylon*. New Haven, Conn.: Yale University Press.

Ptolemy, Claudius. 1460/1991. *The Geography*. Translated and edited by Edward Luther Stevenson. Introduction by Joseph Fischer. New York: Dover; reprint of New York Public Library, 1932.

Rabelais, François, ed. 1534. *Topographia antiquae Romae*, by Bartolomeo Marliani. Lyons: Sebastian Gryphius.

———. 1994. *Oeuvres complètes*. Edited by Mireille Huchon. Paris: Gallimard/ Pléiade.

Randall, Michael. 1996. *Building Resemblance: Analogical Imagery in Early Renaissance Literature*. Baltimore: The Johns Hopkins University Press.

————. 2008. *The Gargantuan Polity: On the Individual and the Community in the French Renaissance.* Toronto: University of Toronto Press.

Riegl, Alois. 1901/1985. *The Late Roman Art Industry.* Translated and with a foreword by Rolfe Winkes. Rome: Giorgio Brettschneider.

Rigolot, François. 2002a. *L'erreur de la Renaissance: Perspectives littéraires.* Paris: Champion.

————. 2002b. *Poésie et Renaissance.* Paris: Seuil.

————. 2008. "Maniérisme et anti-maniérisme dans la rhétorique et la poétique de l'erreur." In *Rhétorique et littérature en Europe de la fin du Moyen Age au XVIIe siècle,* edited by Dominique de Courcelles, 87–101. Turnhout (Belgium): Brepols.

Risset, Jacqueline. 1971. *L'anagramme du désir: Essai sur la "Délie" de Maurice Scève.* Rome: Mario Bulzoni Editore.

Rondot, Nathalis. 1897. *Bernard Salomon: Peintre et tailleur d'histoire à Lyon, au XVIe siècle.* Lyons: Mougin-Rusand.

Ronsard, Pierre de. 1552. *Les Amours de P. de Ronsard Vandomoys; ensemble Le cinquiesme de ses odes.* Paris: La Veufve M. de La Porte.

————. 1553. *Les Amours de P. de Ronsard; nouvellement augmentées par lui & commentées par Marc Antoine du Muret. Plus quelques odes de l'auteur, non encor imprimées.* Paris: La Veuve M. de la Porte.

————. 1559. *Le second livre des Meslanges de Pierre de Ronsard vandomoys.* Paris: Vincent Sertenas.

————. 1560. *Oeuvres de P. de Ronsard, gentil-homme vandomois.* 4 vols. in 3. Paris: G. Buon.

————. 1562. *Discours des miseres de ce temps.* Paris: Buon.

————. 1563. *Réponse aux injures et calomnies de je ne sçay quels predicanteraux et ministraux de Genève.* Paris: Buon.

————. 1914–67. *Oeuvres complètes.* Critical edition by Paul Laumonier et al. 20 vols. Paris: STFM. [Vols. 1–5, Hachette; vols. 6–10, Droz; vols. 11–20, Didier]

————. 1963. *Les "Amours."* Edited by Henri Weber and Catherine Weber. Paris: Garnier Frères.

————. 1993 and 1994. *Oeuvres complètes.* 2 vols. Edited by Jean Céard, Daniel Ménager, and Michel Simonin. Paris: Gallimard/Pléiade.

Ronsin, Albert. 1991. *La fortune d'un nom: Le baptême du nouveau monde à Saint-Dié-des-Vosges: "Cosmographiae Introductio" suivi des lettres d'Amerigo Vespucci.* Lyons: Jérôme Millon.

Rosolato, Guy. 1996. *La portée du désir ou la psychanalyse même.* Paris: Presses Universitaires de France.

Saenger, Paul. 1997. *Space between Words: The Origins of Silent Reading.* Stanford, Calif.: Stanford University Press.

Sand, George. 1846. *La mare au diable.* 2 vols. Paris: Desessart.

Saulnier, Verdun-L. 1948–49. *Maurice Scève.* 2 vols. Paris: Klincksieck.

Saunders, Alison, ed. 1993. Guillaume de La Perrière, *Le théâtre des bons engines* (circa 1539) and *La Morosophie* (Lyons, 1553). Aldershot: Scolar Press.

Scève, Maurice. 1544. *Délie: Obiect de plus haulte vertu.* Lyons: Chez Sulpice Sabon, pour Antoine Constantin.

————. 1544/1980. *Délie: Obiect de plus haulte vertu.* Critical edition by Ian McFar-
lane. Cambridge: Cambridge University Press.

————. 1544/2004. *Délie: Obiect de plus haute vertu.* Critical edition by Gérard
Defaux. 2 vols. Geneva: Droz. [See also Defaux 2004.]

————. 1544/2000. *Délie: Obiect de plus haute vertu.* Edited by Cécile Alduy and with
an introduction and bibliography. Paris: Société des Textes Français Modernes.

————. 1547. *Saulsaye.* Lyons: Jean de Tournes. [Houghton Library, Harvard University Typ 515.47.772]

————. 1564. *Microcosme.* Lyons: Jean de Tournes.

Schapiro, Meyer. 1973. *Words and Pictures: On the Literal and the Symbolic in the
Illustration of a Text.* The Hague: Mouton.

————. 1996. *Words, Script, and Pictures: Semiotics of Visual Language.* New York:
Braziller.

Schmidt, Albert-Marie, ed. 1953. *Poètes du XVIe siècle.* Paris: Gallimard/Pléiade.

Schulz, Juergen. 1978. "Jacobo de'Barbari's View of Venice: Map Making, City
Views, and Moralized Geography before 1500." *Art Bulletin* 60: 425–74.

Sharratt, Peter. 2005. *Bernard Salomon: Illustrateur Lyonnais.* Geneva: Droz.

Shirley, Rodney. 1987. *The Mapping of the World: Early Printed World Maps, 1472–1700.*
London: Holland Press.

Skelton, Raleigh A. 1966. Introduction to the Facsimile of the 1540 edition of Münster's edition (Basel, 1540) of Ptolemy, *Geographia.* 3rd. ser., vol. 5. Amsterdam:
Theatrum Orbis Terrarum.

Stephens, Walter. 1989. *Giants in Those Days: Folklore, Ancient History, and National-
ism.* Lincoln: University of Nebraska Press.

Stevenson, Edward Luther, ed. 1932/1991. *Claudius Ptolemy: The Geography.* New
York: Dover.

Symeone, Gabriele. 1560. *Dialogo pio et speculativo, con diverse sentenze Latine &
volgari.* Lyons: Guillaume Roville.

Terreaux, Louis. 1968. *Ronsard, correcteur de ses oeuvres: Les variantes des Odes et des
deux premiers livres des "Amours."* Geneva: Droz.

Topsell, Edward. 1658/1967. *History of Four Footed Beasts and Serpents.* Cambridge,
Mass.: Da Capo Press.

Tory, Geoffroy, ed. 1512. *Itinerarium provinciarum omnium Antonioni Augusti.* Paris:
Henri Estienne.

————. 1529. *Champ fleury: Auquel est contenu Lart & Science de la deue & vraye
Proportion des lettres attiques, quo[n] dit autrement Lettres Antiques, & vulgaire-
ment Lettres Romaines proportionnees selon le Corps & Visage humain . . .* Paris:
Geoffroy Tory.

Turner, Henry. 2006. *The English Renaissance Stage: Geometry, Poetics, and the Practical Spatial Arts, 1580–1630.* Oxford: Oxford University Press.

Waldseemüller, Martin. 1507. *Cosmographiae introductio: Cum quibusdam geometriae
ac astronomiae principiis ad eam rem necessariis; Insuper quattuor Americi Vespucij
navigationes; Uniuersalis Cosmographiae descriptio tam . . .* Deodate: n.p.

Weber, Henri. 1955. *La création poétique au XVIe siècle en France: De Maurice Scève à
Agrippa d'Aubigné.* Paris: Nizet.

Weber, Henri, and Catherine Weber, eds. 1963/1998. Critical edition of Pierre de Ronsard, Les "Amours." With an updated bibliography. Paris: Editions Garnier. [See also Ronsard.]

Wey Gómez, Nicolás. 2008. The Tropics of Empire: Why Columbus Sailed South to the Indies. Cambridge, Mass.: MIT Press.

Williams, Peter. 2009. Snail. London: Reaktion Books.

Wood, Christopher S. 2008. Forgery Replica Fiction: Temporalities of German Renaissance Art. Chicago: University of Chicago Press.

Woodward, David, ed. 2007. The History of Cartography. Vol. 3, The European Renaissance. Chicago: University of Chicago Press.

Zerner, Henri. 1996. L'invention du classicisme: L'art français de la Renaissance. Paris: Flammarion.

Zumthor, Paul. 1993. La mesure du monde. Paris: Seuil.

Index

Abraham, Nicolas, 212n9

Alberti, Leon Battista, 217n7

Alciati, Andreas, 9, 18, 122, 123, 153; definition of emblem, 87

Alduy, Cécile, 119, 152, 211n24

alterity, 16–18, 20, 23, 25, 33, 60, 117, 167, 202–4; in "Apologie de Raimond Sebond," 194–96; embrace of, in *Pantagruel,* 31–40, 50; fear in *Pantagruel,* 38, 49; in Ronsard, 150, 173; in Scève, 132; topography in, 51

Antonioli, Robert, 213n12

Apian, Pieter, 8, 12–17, 25, 51, 53–79, 80, 82, 83, 88, 126, 177, 197, 201–6, 210n13, 210n15; biography of, 216n6; city view in, 14–15, 17; and contemplation, 67; *Cosmographia* and new world, 56, 60, 69; and *Cosmographiae introductio* (Venice, 1532), 69, 214n24, 216n6, 218n18; and *Cosmographia Petri Apiani* (Antwerp, 1550), 217nn10–11; and *Cosmographia Petri Apiani* (Paris, 1553), 217nn10–11, 217–18n13; and *Cosmographie . . . , libvre tresutile* (Antwerp, 1544), 217n6, 217n8, 217n12;

description of places in, 225nn1–2; *dimensio pedalis* versus *dimensio manualis,* 54; and dismemberment of, 203; emblematic disposition of images in, 18, 70; and Gemma Frisius, 57; and mensuration, 53–54, 57, 73; monocular and binocular vision in, 13; Parisian editions (*Cosmographie de Pierre Apian,* 1551 and 1553), 12, 55–56, 59–69, 71–74, 215n1, 217n8, 218n15; prefatory matter of, 57–61; presence of human body in, 77–78; relation of unknown in, 56; and Scève, 222n11; similitude of geography and topography, 12–16; *speculum cosmographicum* or cosmographical mirror in, 73–75, 202, 205; tension of cosmography and topography in, 14, 15, 55, 58–59, 61, 72–79; title page of, 62; topographical impulse in, 56; volvelles in, 57, 73–74; world map of 1520 (after Waldseemüller), 76, 218n20. *See also* Frisius, Gemma

Aristotle, 65, 227n14

ars moriendi, 6

Tom Conley is Abbott Lawrence Lowell Professor in the departments of Romance Languages and Visual and Environmental Studies at Harvard University. He is the author of *Film Hieroglyphs: Ruptures in Classical Cinema* (Minnesota, 1991 and 2006), *The Graphic Unconscious in Early Modern French Writing, The Self-Made Map: Cartographic Writing in Early Modern France* (Minnesota, 1996 and 2010), and *Cartographic Cinema* (Minnesota, 2007).